Drying of Porous Materials

Drying of Porous Materials

Edited by

Stefan Jan Kowalski
Poznań University of Technology, Poland

Reprinted from *Transport in Porous Media*, Volume 66(1–2), 2007

A C.I.P. Catalogue record for this book is available from the library of Congress.

ISBN-978-1-4020-5479-2 (HB)
ISBN-978-1-4020-5480-8 (e-book)

Published by Springer,
P.O. Box 17, 3300 AA Dordrecht, The Netherlands.

www.springer.com

Cover Illustration (background)
Photo of wood sample after drying (photograph taken by Andrzej Rybicki)

Printed on acid-free paper

All Rights Reserved
© 2007 Springer
No part of this work may be reproduced, stored in a retrieval system, or transmitted
in any form or by any means, electronic, mechanical, photocopying, microfilming, recording
or otherwise, without written permission from the Publisher, with the exception
of any material supplied specifically for the purpose of being entered
and executed on a computer system, for exclusive use by the purchaser of the work.

Printed in the Netherlands.

CONTENTS

STEFAN JAN KOWALSKI / Preface — 1

ARUN S. MUJUMDAR / An Overview of Innovation in Industrial Drying: Current Status and R&D Needs — 3

ZDZISŁAW PAKOWSKI / Modern Methods of Drying Nanomaterials — 19

YOSHINORI ITAYA, SHIGERU UCHIYAMA and SHIGEKATSU MORI / Internal Heating Effect and Enhancement of Drying of Ceramics by Microwave Heating with Dynamic Control — 29

STEFAN JAN KOWALSKI and ANDRZEJ RYBICKI / The Vapour–Liquid Interface and Stresses in Dried Bodies — 43

PATRICK PERRÉ / Multiscale Aspects of Heat and Mass Transfer During Drying — 59

STANISŁAW PABIS / Theoretical Models of Vegetable Drying by Convection — 77

MARCIN PIATKOWSKI and IRENEUSZ ZBICINSKI / Analysis of the Mechanism of Counter-Current Spray Drying — 89

THOMAS METZGER, MARZENA KWAPINSKA, MIRKO PEGLOW, GABRIELA SAAGE and EVANGELOS TSOTSAS / Modern Modelling Methods in Drying — 103

MUSIELAK GRZEGORZ and BANASZAK JACEK / Non-linear Heat and Mass Transfer During Convective Drying of Kaolin Cylinder under Non-steady Conditions — 121

WIESLAW OLEK and JERZY WERES / Effects of the Method of Identification of the Diffusion Coefficient on Accuracy of Modeling Bound Water Transfer in Wood — 135

STEFAN JAN KOWALSKI and ANNA SMOCZKIEWICZ-WOJCIECHOWSKA / Stresses in Dried Wood. Modelling and Experimental Identification — 145

JAN STAWCZYK, SHENG LI, DOROTA WITROWA-RAJCHERT and ANNA FABISIAK / Kinetics of Atmospheric Freeze-Drying of Apple — 159

MICHAL ARASZKIEWICZ, ANTONI KOZIOL, ANITA LUPINSKA and MICHAL LUPINSKI / Microwave Drying of Various Shape Particles Suspended in an Air Stream — 173

ROMAN G. SZAFRAN and ANDRZEJ KMIEC / Periodic Fluctuations of Flow and Porosity in Spouted Beds — 187

J.F. NASTAJ and B. AMBROŻEK / Modeling of Vacuum Desorption of
   Multicomponent Moisture in Freeze Drying   201
ARTUR POŚWIATA and ZBIGNIEW SZWAST / Optimization of Fine Solid
   Drying in Bubble Fluidized Bed   219
STEFAN JAN KOWALSKI / Referees   233

Preface

Stefan Jan Kowalski

Published online: 10 October 2006
© Springer Science+Business Media B.V. 2006

I am very pleased to act as Guest Editor for this Special Issue of the *International Journal Transport in Porous Media*, entitled **Drying of Porous Materials**. This Special Issue consists of selected papers presented at the XI Polish Drying Symposium, held in Poznań, a historic and beautiful city in Western Poland, on September 13–16, 2005. The Symposium, organized jointly by the Drying Section of the Committee of Chemical and Process Engineering of the Polish Academy of Sciences, the Polish Drying Association, Poznań University of Technology and August Cieszkowski Agricultural University of Poznań, attracted 94 participants, including foreign guests from 13 countries. The Polish Drying Symposium series was initiated in 1972 by Professor Czesław Stumiłło, the Honorable President of this Symposium.

The Symposium language was English, both for lectures and poster presentations. This allowed extended foreign participation and a fruitful exchange of ideas between Polish and international participants. Several prominent scientists were present, including, among others, A.S. Mujumdar, P. Perre, Y. Itaya, T. Kudra, G. Srzednicki, A.E. Djomeh, O. Yaldiz, T. Metzger from abroad and Cz. Strumiłło, Z. Pakowski, I. Zbiciński, S. Pabis, P.P. Lewicki, A. Kmieć from Poland. Altogether, 34 lectures and 7 excellent keynote lectures were presented and the poster session accommodated 43 posters.

The subject of the Symposium has been traditionally shared by agri/food (ca. 35%) and all other drying subjects. Among agri/food subjects, the drying of wood was strongly represented and a considerable number of papers covered drying process simulation and parameter identification. A brief survey of all presentations indicates that there is no national bias or specialty characterizing Polish drying research. Instead, it can be noticed that it encompasses a majority of topics represented at all drying conferences, including the well known IDS (International Drying Symposium).

S. J. Kowalski (✉)
Poznan University of Technology, Institute of Technology and Chemical Engineering,
pl. Marii Sklodowskiej Curie 2, 60–965 Poznan,
Poland
e-mail: stefan.j.kowalski@put.poznan.pl

This indicates that Polish drying research closely follows global cutting edge research on subjects of drying, including such emerging areas as CFD (Computational Fluid Dynamics) application, drying of nanomaterials, microencapsulation, supercritical drying, etc.

The papers selected for inclusion in this special issue concern mostly drying of liquid-saturated porous materials, but not exclusively. The paper by A.S. Mujumdar "An Overview of Global R&D in Industrial Drying: Current Status and Future" presents a review of the current problems concerning research and development on drying, and particularly on industrial drying. Several papers deal with spray drying, or drying in a bubbling fluidized bed, whose modeling is quite similar to that encountered in the theory of the liquid-saturated porous media.

The Editor of TIPM kindly agreed to publish this special issue of the Journal, with several unpublished papers reflecting some aspects of the theory of saturated porous media applied to heat and mass transfer in drying, as well as an analysis of drying induced stresses. Out of 84 papers presented at the Symposium, the Scientific Committee and the specialists on porous materials, both from Poland and abroad, who acted as the reviewers, have selected 16 papers for this Special Issue. A number of original papers concerning drying, but not necessarily drying of porous materials, presented at the Symposium, will be published in a special issue of *Drying Technology–an International Journal* as well as in a regular issue of the journal *Chemical and Process Engineering*.

As the Guest Editor, I would like to take this opportunity to express my cordial thanks to Prof. Jerzy Weres, the co-chairman of the Symposium and his co-workers from the August Cieszkowski Agricultural University in Poznań, for their help in organizing the Symposium. I also address my sincere thanks to my co-workers from the Department of Process Engineering of the Institute of Technology and Chemical Engineering, Poznań University of Technology for their efforts and precious help in all aspects of the Symposium's organization.

Financial support for the Symposium from the following sponsors is gratefully acknowledged:

- Rector of Poznań University of Technology,
- Ministry of National Education and Sport,
- Foundation on Behalf of Development of Poznań University of Technology,
- International Tobacco Machinery Poland Ltd. in Radom.

Transp Porous Med (2007) 66:3–18
DOI 10.1007/s11242-006-9018-y

ORIGINAL PAPER

An overview of innovation in industrial drying: current status and R&D needs

Arun S. Mujumdar

Received: 30 November 2005 / Accepted: 26 March 2006 /
Published online: 30 August 2006
© Springer Science+Business Media B.V. 2006

Abstract Over the past three decades there has been nearly exponential growth in drying R&D on a global scale. Although thermal drying had always been the workhorse of almost all major industrial sectors, the need for and opportunities in basic as well as industrial research became clear only after the energy crisis of the early 1970s. Although the price of oil did drop subsequently the awareness of the significance of improving the drying operation to save energy, improve product quality as well as reduce environmental effect remained and indeed has flourished over recent years. New drying technologies, better operational strategies and control of industrial dryers, as well as improved and more reliable scale-up methodologies have contributed to better cost-effectiveness and better quality dried products. Yet there is no universally or even widely applicable drying theory on the horizon. Most mathematical models of drying remain product-equipment specific for a variety of reasons. In this paper, we examine the role of innovation in drying in various industrial sectors, e.g. paper, wood, foods, agriculture, waste management, etc. Progress made over the past three decades and the challenges ahead are outlined. Some areas in need of further research are identified. Examples of intensification of innovation in dryer designs via mathematical modeling are discussed. Finally, the need for closer interaction between academia and industry is stressed as the key to successful drying R&D in the coming decade.

Keywords Innovation · Intensification of innovation · Mathematical models · Energy aspects · Environmental impact · Novel dryers · Research and development needs

A. S. Mujumdar (✉)
Department of Mechanical Engineering,
National University of Singapore,
9 Engineering Drive 1, 117576 Singapore, Singapore
e-mail: mpeasm@nus.edu.sg

1 Introduction

Thermal drying has been recognized as an important unit operation as it is energy-intensive and has a decisive effect on the quality of most products that are dried commercially. Escalating energy costs, demand for eco-friendly and sustainable technologies as well as the rising consumer demand for higher quality products, have given fresh incentives to industry and academia to devote great effort to drying R&D. Fortunately, this area does not demand a massive perfusion of R&D funds to come up with valuable insights and innovations, with only a few exceptions. Indeed, there is already a sustainable level of R&D support—both in terms of human and financial resources—around the world. Emerging economies of the world, such as China, Brazil, India, etc, have picked up the slack caused by the fully developed economies of the west moving towards nanotechnologies. Overall, the global R&D effort has been rising despite precipitous drops in North America and Europe. With 12–25% of the national industrial energy consumption attributable to industrial drying in developed countries, it is only a matter of time before high energy costs will stimulate further R&D in drying.

The fact that tens of thousands of products need to be dried in over a 100 variants of dryers provides major and ample opportunities for innovation. In this paper, we start with the definition of innovation and how it may be intensified. This is followed by a discussion of some new technologies in comparison with the traditional ones, which still dominate the market. The need for industry–academia interaction and a pro-active role of industry—the ultimate beneficiary of all R&D regardless of its origin—is stressed.

2 Need for and role of R&D

Researchers in academia as well as industry along with granting agencies consistently agree on the need for more R&D funds. They argue that R&D funds should be rightfully considered as investment rather than expenditure. In either case, it is necessary to account for the outlays on R&D in terms of its economic and/or social benefits. It is essential to look critically at the cost/benefit ratio for R&D funds in general and provide appropriate justification for such expenditures or investments. This is more readily—not necessarily easily or reliably—achieved in the business or industrial world. When public funds are used for R&D—a major source for most nations—it is a much more difficult task.

There is much scholarly literature on the economic returns on publicly funded basic research (e.g. A.J. Salter and Ben R. Martin, Research Policy, Vol. 30, 2001, pp. 509–532). As these authors point out, research output may be information or knowledge that can be used to economic advantage; they postulate that much of the publicly funded R&D output is of informational nature and the knowledge created is "non-rival" and "non-excludable". Non-rival knowledge is defined as that which others can use "without detracting from the knowledge of the producers". Non-excludable implies that no one can be stopped from using this knowledge—even competitors have free access to it although they did not pay for it directly. This is also the nature of information and knowledge disseminated by journals such as Drying Technology.

Utilization of "free" informational knowledge requires significant investment to understand and use it to advantage. Thus, scientific knowledge is really not available "freely" but only to those who have the necessary expertise to access it. An OECD

Report (1996) states: "Knowledge and information abound; it is the capacity to use it that is scarce". Information is available to all, but only those with the right capabilities can convert it to knowledge and use it to innovate.

I postulate that the rate of technological innovation depends directly on the rate of generation of informational knowledge and the effectiveness of its utilization; the latter is a measure of the ability to assimilate or exploit the knowledge. Efficient dissemination of knowledge is important but it is equally important to develop the ability to utilize it. Academic institutions are responsible for developing such ability. If they can also make a valuable contribution to the generation of new knowledge, then they are very effective in enhancing the rate of innovation, which drives the economic growth of nations.

Talking about sustainable development is in vogue these days. Clearly, it makes a lot of sense and the world will be a better place if all development were truly sustainable. I believe that this concept is applicable to research and development effort as well, be it in academia or in industry.

We are concerned about the continually shrinking R&D funding pie almost all over the world. Granting bodies have tried to cut the pie in many different ways. Usually, areas that are currently popular or fashionable receive larger shares of the pie, thus reducing the funding for some other key or core areas—or worse, even eliminating funding for many of the so-called traditional areas of research. The implication is that enough R&D has already been conducted in areas that have existed for longer periods. New is automatically assumed to be innovative, creative and thus valuable for the future development of the economy. Larger portions of R&D funding have gone into energy technologies, environmental issues, bio-technology, IT, etc. over the past two decades, and more recently into nano-science and nano-technology. Clearly, all these areas are important and deserve funding. The issue is how much and at what cost to the other areas. Will it cause some core incompetencies in future?

This is where I think, we need to examine my idea of *sustainable research* level. Just for illustration, if all institutions around the world focus their effort on developing nano-technology, are there enough opportunities for so many to make original contributions? Also, is the market large enough for all the players to obtain adequate returns on their investments? Some institutions will lead and outpace most others as a result of their access to higher levels of human and financial resources. At the same time, areas that industry currently needs and technologies that form the lifeline of current businesses will be penalized as no new funding is made available for the new R&D areas. Overcrowding of pre-selected research areas is as risky as under-populating them with under-funding. Over-funding does not assure development of innovative ideas; it may even impede it as funding becomes easier to obtain and hence non-competitive. Also, projections of potential major windfall from new technologies are fraught with uncertainties.

I wish to postulate that for each research field at any given time and any given location, there is a sustainable level of R&D funding support beyond which the returns on investment will necessarily decline. The opportunity cost of not doing R&D is other areas will rise as well. Drying is considered a mature area: in many ways it is. However, there are still many unsolved, complex problems that deserve attention and offer challenges to researchers. The return on the modest levels of R&D funding needed to carry out drying research can be substantial since this operation is so energy-intensive and has a direct impact on product quality and the environment. What is needed is a focused effort with collaboration between industry and academia.

Aside from producing highly qualified researchers, such cooperation will also produce improved technologies that will benefit industry and the consumer at large. It is obviously unlikely that industries will be operating without the unit operation of drying, which means investment in drying R&D will have a useful pay-off at all times. One valuable effect of globalization and free flow of research results is that not all countries need to be involved in many areas of scientific R&D that can be best left to countries with needed resources.

3 Innovation

According to Howard and Guile (1992) innovation is defined as follows:

"A process that begins with an invention, proceeds with development of the invention, and results in the introduction of new product, process or service in the marketplace".

To make it into a free marketplace, the innovation must be cost-effective. What are the motivating factors for innovation? For drying technologies, I offer the following checklist; one or more of the following attributes may call for an innovative replacement of existing products, operation or process:

- New product or process not made or invented heretofore;
- Higher capacities than current technology permits;
- Better quality and quality control than currently feasible;
- Reduced environmental impact;
- Safer operation;
- Better efficiency (resulting in lower cost);
- Lower cost (overall, i.e., lower investment and running costs).

Innovation is crucial to the survival of industries with short time scales (or life cycles) of products/processes, i.e., a short half-life (less than one year, as in the case of some electronic and computer products). For longer half-lives (e.g., 10–20 years typical of drying technologies) innovations come slowly and are less readily accepted. The need for replacement of current hardware with newer and better hardware is less frequent and the payback is less attractive. It is important to note that newness per se is not good enough justification for adoption of "innovative" technologies.

Innovations may be revolutionary or evolutionary. Evolutionary innovations, often based on adaptive designs, have shorter gestation periods, shorter times for market acceptance and are typically a result of "market-pull," something the marketplace demands, i.e., a need exists currently for the product or process. These usually result from a linear model of the innovation process (an intelligent modification of the dominant design is an example). Revolutionary innovations, on the other hand, are few and far between, have longer gestation periods, may have larger market resistance and are often a result of "technology-push," where development of a new technology elsewhere prompts design of a new product or process for which market demand may have to be created. They are riskier and often require larger R&D expenditures, as well as sustained marketing efforts. The time from concept to market can be very long for some new technologies. It is well known that the concept of a helicopter appeared some 500 years before the first helicopter took to the air. The idea of using superheated steam as the drying medium was well publicized over 100 years ago, yet its real commercial potential was first realized only about 50 years ago and that too not fully. In fact, it is not fully understood even today! The most recent example of this long-gestation period is the Condebelt drying process for high basis weight (thick grade) paperboard, proposed and developed by the late Dr. Jukka Lehtinen for Valmet Oy of Finland. It took a full 20 years of patience and high-quality R&D before the process was first deployed successfully.

It is natural to inquire if it is possible to "guesstimate" the best time when the marketplace requires an innovative technology or the mature technology of the day is ripe for replacement. Foster's well-known "S" curve (Foster 1986), which gives a sigmoid relationship between product or process performance indicators and resources devoted to develop the corresponding technology, is a valuable tool for such tasks. Every technology has its asymptotic limit of performance. When this happens (or even sooner), the time is ripe to look for alternate technologies, which should not be incremental improvements on the dominant design but truly new concepts, which once developed to their full potential, will yield a performance level well above that of the current one.

Table 1 lists examples of some new drying technologies that were developed via technology-push versus market-pull. In some cases, a sharp distribution or grouping in just two types is not possible since a "market-pulled" development may require a "technology-push" to succeed. For example, development of new materials was key to successful implementation of the Condebelt or impulse drying process for paper.

Innovation has become a buzzword in academia and industry alike. Drying R&D and technology is no exception. We have been promoting innovation in drying for over two decades via the IDS series as well as Drying Technology journal. However, the quantitative measurement of innovative performance remains an elusive task. There are no widely accepted indicators of innovative performance or a common set of indicators to assess the returns on investment. Such an indicator or set of indicators is crucial to managing innovation. Some literature studies have used R&D inputs, number of patents, number of patent citations, counts of new product launches, etc., as indicators of innovative performance in industry. The task is harder for academic institutions, however, which often remain fixated on the impact of the "printed matter" generated rather than industrial significance, which is hard to measure.

Table 1 Examples of new drying technologies developed through technology-push and market-pull

Technology push[a]	Market-Pull[b]
Microwave/RF/induction/ultrasonic drying	Superheated steam dryers— enhanced energy efficiency, better quality product, reduced environmental impact, safety, etc.
Heatpump dryers	
Pulse combustion drying—PC developed for propulsion and later for combustion applications	Impulse drying/Condebelt drying of paper (also needed technology-push to succeed)
Vibrating bed dryers—originally developed for solids conveying	Combined spray-fluid bed dryers—to improve economics of spray drying
Impinging streams (opposing jets)—originally developed for mixing, combustion applications	Intermittent drying—enhance

[a] Technology originally developed for other applications applied to drying; also may be "science-push" type

[b] Developed to meet current or future market demand

It is always interesting to look at Nature for truly creative ways of solving complex problems. A recent article in Mechanical Engineering Design (2004) discussed a species of beetle, called the Bombardier beetle, which squirts its predators with a high-pressure pulsed spray of a boiling hot toxic liquid. The chemistry of the liquid and the mechanism of the pulsed spray have been studied in depth by biologists and biochemists for over two decades. Research by Professor G. Eisner of Cornell University discovered that the Bombardier beetle produces hydrogen peroxidase and hydroquinone, and when attacked by a predator, it can mix the two in a tiny heart-shaped combustion chamber to produce benzoquinone and steam; the mixture is then emitted as a pulsed jet at temperatures of the order of 100°C. Recent research at the University of Leeds by Professor McIntosh has already found that the unique shape of the beetle's reaction chamber is critically important in maximizing the mass of ejected spray for each "explosion", which can occur about 300 times/s. The shape of the nozzle, which can swivel in any direction, is also important. An in-depth study of this unique creature is expected to yield a solution to the occasional but serious problem of re-igniting a gas turbine aircraft engine, which has cut out at high altitudes and extremely low temperatures. Clearly, the study of natural engineering marvels can help us arrive at novel engineering solutions to complex problems.

Copying such natural mechanisms is a feature of the field of biomimetics in which scientists and engineers learn from the intricate design ideas that nature uses. Indeed, the pulsed combustion-based self-defense mechanism of the Bombardier beetle is an extremely complex design. Such a study will require sophisticated research techniques and multi-disciplinary teams involving biologists, biochemists, chemists as well as engineers. I believe that improved design of the pulsed combustion process could also lead to improved design of novel pulsed combustion dryers to produce powders from liquids. The jury is still out on the viability of this concept.

Revolutionary innovations in any technology are always met with skepticism and even disdain by industry. Everyone wishes to work within their comfort zone. Most industries are risk-averse. Hence, true innovations are hard to market and gain acceptance by industry. However, when they do cross the barrier, they can be truly disruptive in that they have the potential to displace or even supplant the conventional technology of the day. It is noteworthy that in the business world most want to be "second" and not the "first with innovation".

Innovations can thrive only under appropriate incubation conditions. For example, the USA is well recognized as the greatest engine of innovation. It is hard to duplicate it elsewhere with equal success since it is a product of numerous factors, ranging from freedom of thought and expression, stress on independent thinking, ready acceptance of diverse ideas and cultures, immigration of new minds and mindsets, developed financial markets, and risk-taking culture. It is not surprising that the US system is unmatched in recent decades when it comes to bringing innovative ideas and concepts to the world markets. Although there is much hue and cry in the USA about job losses due to outsourcing to the developing world—another innovation from corporate America—it is unlikely it will have a long-term undesirable effect on the US economy since this change will soon precipitate another innovation in business models.

Meteorologist Edward Lorenz in 1972 published a paper entitled "Predictability: Does the flap of a butterfly's wings in Brazil set off a Tornado in Texas", which essentially sums up the Chaos Theory. In complex non-linear systems small perturbations can lead to major disturbances; apparently random events (like flapping of wings and tornado) may actually follow some underlying rules. True innovations follow a non-linear pathway. Thus, there is high likelihood that even minor modifications in dryer designs may eventually lead to major improvements in drying technologies, which we cannot anticipate today. At least, we hope developments in drying are chaotic at least for this reason!

Finally, innovation is not necessarily based on new ideas. Some old ideas will re-emerge, e.g., use of superheated steam drying. It is important to re-visit some old ideas to see if they are viable at the present time. Novel ideas ahead of their time sometimes fizzle out but re-emerge when the time is right.

4 Intensification of innovation

Dodgson et al. (2002) have argued that the innovation process can be enhanced by applying digital technologies, which can simulate, model, integrate and intensify the innovation process via a cost-effective effort. They propose that automation of innovation is feasible. In fact, this is called "Rothwell's concept of the fifth generation" innovation process. Basically, the digital computing power provides a new "*electronic tool kit*" that facilitates transfer, transformation and control of various kinds of information that is required for the successful introduction of innovative products and/or processes in the market place.

The origin of innovation in drying technologies could be a result of: (a) serendipity (chance); (b) fundamental principles of heat and mass transfer, or (c) empiricism. Empirically generated innovations are evolutionary in nature; they are based on incremental improvements of prior technology. Innovations arising from serendipity or fundamentals can be evolutionary or radical. Digital enhancement of innovation is clearly possible primarily with the help of fundamental principles, which can be modeled either deterministically or stochastically with reliable mathematical relationships. As examples, we can cite innovations in spray dryers, flash dryers, high-temperature impinging jet dryers for tissue paper, etc., which allow computer-aided design and modeling. New spray dryer chamber designs can be evaluated with minimal expense and risk using computational fluid dynamic simulations. Both the time from concept to product in the market as well as the cost of the design process can be reduced very significantly via computer simulations. This is not very different from what is already being done in the aircraft industry, e.g., the Boeing 777 was designed primarily by computer-aided simulation and design, unlike its predecessor models. However, transport phenomena occurring in the spray dryer chamber are far more complex than the aerodynamics of flow over the airplane.

Without going into the details of the types and nature of innovation, we can make the following general observations about innovation in the field of drying of solids and liquids:

- Most new dryer design improvements are incremental in nature, e.g., two or three-stage spray drying.
- They are based on intelligent combinations of established technologies, e.g., two-stage spray and fluid bed dryers, steam-tube rotary dryers, ultrasonic spray dryers, etc.
- No disruptive drying technologies (i.e., ones which have supplanted traditional technologies) have appeared on the horizon as yet.

- Truly novel technologies, which significantly from the conventional ones, are not readily accepted by industry, e.g., superheated steam impinging jet drying of paper, Condebelt drying of liner board, pulse combustion dryers, use of a bath of liquid metal to dry paper, the Remaflam process for textile drying, impinging stream drying for sludge, etc.
- The need for new drying hardware is typically limited due to the long life-cycle of drying equipment; for example, most dryers have a life span of 20–40 years. Hence, the need for replacement with new equipment is limited.
- Often, firms which are first to commercialize a new product or process in the market, do not necessarily benefit from being the true innovators. This phenomenon hinders the introduction of new technologies. A fast second or even a slow third might outperform the innovator. According to Teece (1986), this observation is particularly pertinent to science and engineering-based companies that have the illusion that development of new products that meet customer needs will ensure success. A classical example of this phenomenon is RC Cola, which was first to introduce Cola in a can; however, it was Coca Cola and Pepsi Cola that dominated the market. There are numerous such examples, e.g., pocket calculator (Bowmar was outperformed by Texas Instruments, HP and others); personal computer (Xerox outperformed by Apple); jet aircraft (de Havilland Comet outperformed by Boeing 707), etc. In all these cases, the innovator was first to the market but could not sustain or even attain prominence in the market. Hence, the reluctance to be first in the market with a new process or product.
- Often, innovative concepts are initiated by academic researchers and published without filing for intellectual rights protection in the open literature. Although this is really impactful R&D, little credit accrues to the academic since no archival papers result from industrial use and hence no "citations"! This is a double-edged sword from this author's experience. It is good in that the ideas are widely and freely disseminated for wider economic benefits to society at large. On the other hand, potential industrial interest is dampened by the fact that the innovation is in the public domain so that further R&D investment by industry may not have a payback. The problem of IP (intellectual rights) must be properly addressed to encourage innovation by academia and its transfer to industry and to value it appropriately.
- Since there is potential to innovate via the route of fundamentals and simulations, close industry-academia interaction is another attractive and cost-effective way to intensify innovation. It represents a true win–win situation. A close inspection of the relevant technical literature shows that many new concepts for dryers originate in academia but are rarely utilized in industry since the necessary development work is beyond the scope of academic research. Without the D in R&D no technology transfer occurs. Without the R, there is no potential for D in the R&D combination. *The importance of R is also inherent in the word drying itself! Without R the field itself cannot survive!*
- Drying, contrary to popular belief, is a knowledge-based technology. The manufacturing technologies needed are often very straightforward and found readily in most of the developing world. What is needed for a cost-effective drying system to be made is "knowledge" and "know-how" to counteract deficiencies of knowledge. Knowledge knows no geopolitical boundaries, as is evidenced by the rapid spread of IT technologies in the developing world. Indeed, the latter in some cases has assumed the world-leading role despite capital shortages. This drying

knowledge/ know-how has the potential to be assimilated readily as was done with IT by the developing countries with strongly educated and motivated human capital. Note that the total US market for capital expenditure on dryers was estimated to be only about $500 million by a contractor for the DOE. It is likely an underestimate but perhaps not a far cry from the real figure. Thus design, operation, and optimization of dryers are the key business needs in this field. Clearly, this requires both basic and applied research.
- Drying R&D to be effective must be cross-disciplinary. Typically, product knowledge and techno-economic aspects are best provided by industry collaborators.
- International collaboration, cooperation, and competition is conducive to intensifying innovation in general. Aside from savings in the use of human and financial resources for R&D it can lead to synergy.

5 Conventional versus new drying technologies

It is difficult to make a sharp distinction between what is conventional and what is really new in drying technologies since most of the newer developments are evolutionary, i.e., based on traditional ones; often the transition is seamless and it is not possible to identify where and when it occurred. The following discussion must therefore be taken within this vagueness inherent to the field itself.

Kudra and Mujumdar (1995) have classified and discussed various novel dryers, ranging from laboratory-scale curiosities (e.g., acoustic drying, drying of slurries by impinging sprays over a hot surface) to pilot-scale demonstrations (e.g., pulse combustion dryers, ultrasonic spray dryers, impinging stream dryers) to full-scale commercial dryers (e.g., pulsed fluid beds, superheated steam fluid bed/flash dryers, rotary dryers with drying air injected into the rolling bed). A full discussion of the truly bewildering variety of non-conventional dryers is beyond the scope of this presentation. The interested reader may refer to the book by Kudra and Mujumdar (2002) for a comprehensive coverage of the numerous new drying concepts and technologies. The Handbook of Industrial Drying (Mujumdar 1995) is also a source of relevant information.

Table 2 summarizes the key features of the newer dryers as compared to those of conventional ones for drying of various physical forms of the wet feed material. Note that the new designs are not necessarily better than the traditional ones for all products, but they do offer some advantages that may make them a better choice in some applications. Some of them are simply intelligent combinations of conventional dryers.

Table 3 compares some key features of the newer or emerging drying technologies with those of the more commonly used conventional techniques. In terms of the sources of energy there is no difference. However, in terms of how this energy is delivered and transferred to the wet solid there are some significant differences.

In the chemical industry, the most common drying application involves production of dry particulates from pumpable liquids (solutions, suspensions, or slurries), thin or thick pastes (including sludge), or granular solids. Spray and drum dryers are used most commonly for such applications. Spray dryers today no longer just convert a pumpable liquid to a powder but can be used to produce "engineered" powders with specific particulate size, as well as structure (e.g., agglomerates, granules, or large mono-sized spherical particles). Personnel safety on and around dryers, prevention of environmental pollution, and emphasis on production of a high-quality product at minimum cost are paramount considerations in the design of spray dryers today. With

Table 2 Conventional versus innovative drying techniques

Feed type	Dryer type	New techniques[a]
Liquid suspension	DrumSpray	Fluid/spouted beds of inerts Spray/fluid bed combination Vacuum belt dryers Pulse combustion dryers
Paste/sludge	Spray Drum Paddle	Spouted bed of inert particles Fluid bed (FB) (with solids back-mixing)Superheated steam dryers
Particles	Rotary Flash Fluidised bed (hot air or combustion gas) Conveyor dryer	Superheated steam FBD Vibrated bed, Ring dryer, Pulsated fluid bed, Jet-zone dryer Impinging streams Yamato rotary dryer
Continuous sheets (coated paper, paper, textiles)	Multi-cylinder contact dryers; Impingement dryers	Combined impingement/ radiation Combined impingement and through dryers Impingement and MW or RF or Radiation dryers

MW microwave, *RF* radio frequency

[a] New dryers do not necessarily offer better techno-economic performance for all products. Many require further R&D and market acceptance to succeed

Table 3 Comparison of conventional versus emerging drying technologies

	Conventional	Emerging trends
Energy (heat source)	Natural gas, Oil Biomass, Solar/wind Electricity (MW/RF) Waste heat	No change yet. Renewal energy sources when fossil fuel becomes very expensive
Fossil fuel combustion	Conventional	Pulse combustion
Mode of heat transfer	Convection (>85%) Conduction Radiation (< 1%) Microwave/radio frequency	Hybrid modes Non-adiabatic dryers Periodic or on/off heat input
Drying medium (convective dryers)	Hot air Flue gases	Superheated steam Hot air + superheated steam mixture or two-stage
Number of stages	One—most common Two or three—same dryer type	Multi-staging with different dryer types
Dryer control	Manual Automatic	Fuzzy logic, Model based control, Artificial neural nets

the help of computer simulations, better designs of the dryer chamber and air flows within the dryer have led to minimal wall deposit problems in spray dryers. A new spray dryer concept even uses a flexible canvas cone instead of the usual metallic one. Horizontal spray chamber layouts are also being examined via CFD simulations.

The development of in-place cleanable bag filters makes it possible to retain the particulates within the dryer chamber; this is achieved by mounting the filter elements in the roof of the spray dryer chamber. No external cyclones are needed in this case. This technology, coupled with the popular fluidized-spray dryer featuring a fluid bed dryer integrated into the base of the spray dryer chamber, allows efficient

production of dust-free agglomerated or granulated products at substantially lower product temperatures than those found in conventional spray dryers.

For drying granular or particulate solids, the most common dryers in use today are cascading rotary dryers with or without internal steam tubes, conveyor dryers, and continuous tray dryers (e.g., turbo or plate dryers), which must compete with fluidized-bed dryers (with or without internal exchangers) and vibrated bed dryers, among others. At least 20 variants exist of the fluidized-bed dryer alone. For larger particles, a spouted bed dryer is preferable to the conventional fluidized-bed dryer. For difficult-to-fluidize, sticky particles or feedstocks with a wide particle size distribution, the vibrated bed dryer offers advantages over the conventional fluid bed because it allows use of low drying air velocities while mechanical vibration assists in pseudo-fluidizing the solids. Recently, the pulsed fluidization technique has found some interesting applications similar to those for vibrated fluid beds.

Fluidized-bed dryers have been operated successfully using superheated steam as the drying medium both at near atmospheric pressures (e.g., drying of pulverized lignite) and at elevated pressures (3–5 bar for drying of beet pulp). In addition to eliminating potential fire and explosion hazards, use of superheated steam permits utilization of the exhaust steam by condensation, reheating, or compression. Of course, such steam is often contaminated and must be cleaned before re-utilization, depending on the application. Net energy consumption in superheated steam dryers may be as low as 700–1000 kJ/kg water evaporated, which is 5–10 times lower than many conventional convective air drying systems consume. Mujumdar (1990) discusses the principles, practice and potential of the rapidly emerging superheated steam drying technologies in various industrial sectors, principally for drying paper, wood, some processed foods, sludge, etc. Numerous relevant references are cited as well. It is commonly recognized as the drying technology of the future. It is known that low-temperature superheated steam dryers are feasible for drying very heat-sensitive materials like fruits and vegetables, silk cocoons, etc. SEM pictures of steam-dried products clearly show the quality advantages such products can offer. Easy to re-hydrate dried fruits such as strawberries, pears, apple, have been produced by low-temperature steam drying in Argentina. Vacuum steam drying of wood is probably the best technology for drying of wood. It is more commonly found in the developing countries (e.g., Malaysia, India, Thailand, etc.) rather than in developed countries of North America which have invested heavily in the hot air kiln dryers for wood, which cannot easily be switched to steam operation.

The use of volumetric heating by microwave (MW) and radio frequency (RF) fields has yet to make major inroads in the chemical industry. It is well established that MW/RF-assisted drying is faster, but the energy costs are also significantly higher and scale-up for large production capacities is much more difficult. Such dryers are expected to find some niche applications. RF drying under vacuum has been applied successfully on a commercial scale by a Canadian company for drying of timber and thick veneer with future applications anticipated for drying of chemicals, polymers, and foods. An RF dryer, in conjunction with impingement with hot air jets, is already a commercial process for drying of coated paper. The efficiency of conversion of line power into electromagnetic energy and the cost of electricity are major impediments in commercializing this technology. Hybrid drying concepts will certainly make rapid inroads in industrial drying in the coming decade. These consist of intelligent combinations of well-established drying technologies and hence involve less risk. They can

combine advantages as well as limitations of each individual technology. Hence, care must be exercised in designing hybrid dryers.

Some of the newer drying technologies utilize newly developed gas–solid contactors as dryers for particulates, e.g., impinging streams, rotating spouted beds, pulsed-fluid beds, etc. In batch drying, one can take advantage of intermittent multi-mode heat input to optimize both the energy consumption and product quality. Intermittencies may be in the form of changes in flow regimes (packed versus fluidized), drying air conditions (air velocity, temperature, humidity, etc), operating pressure (atmospheric, above or below atmospheric), and/or changes in modes of heat input (convection, conduction, radiation or microwave/RF applied sequentially or jointly). It has been shown by many investigators that intermittent operation enhances energy efficiency as well as product quality for heat-sensitive materials. When several parameters of the process are altered with time then, we have multiple intermittencies. This is a field that has not been explored in depth yet, although this can be done quite easily. Many of the traditional dryers, which we assume to be operating under continuous heat input, in fact operate intermittently, e.g., the multi-cylinder paper machine. On the other hand, all batch dryers requiring very long-dwell times in the dryer are often dried under time-varying drying schedules, e.g., freeze dryers, wood drying kilns, etc.

The list in Table 4 is very short and included only for illustrative purposes. The proceedings volumes of IDS series and several other drying conferences (e.g. ADC,NDC, IADC, etc.) and the archival journal Drying Technology (www.dekker.com) provide rich sources for new ideas for further R&D. Searches on the Internet are also invaluable in identifying new technologies and new research challenges. The whole field is far from maturity; it is still in a state of rapid flux.

6 Some R&D needs in drying

It is impossible to summarize R&D needs for all types of industrial dryers. The problems are as diverse as the equipment and often the wet materials processed. Since the prospects for a universally applicable theory of drying is not on the horizon any time soon, scale up of dryers will continue to rely heavily on carefully planned experiments with phenomenological models of limited applicability. Most models are applicable to specific product-equipment combinations (with notable exceptions, of course). Some 60,000 products need to be dried at different scales in over 100 dryer types. The need for R&D is therefore enormous. What is included here covers just a few basic ideas and the interested reader can readily extend them into some innovative designs for improved drying technologies.

It is hard to make an all-inclusive wish list of R&D projects that would be highly worthwhile, in my opinion. The list I made in 1990 for a similar presentation still stands as is and has grown enormously as I became increasingly aware of the industrial needs and intricacies of drying. With advances in digital computing capabilities and innovative analytical instruments, we can now search deeper and examine moisture movement at the microscopic level. Advances in new materials and nanotechnology have made new demands on drying research. In the old days, we were interested only in scale-up—going from smaller to larger scale. Now there is a new need to also go from small to smaller scale! Spray drying of minute quantities of drugs or advanced materials is a case in point.

Control of dryers based on quality measurement in real time is a challenge. For example, an artificial nose can sense flavor quantitatively and trigger appropriate

control action. One can think of smart dryers that automatically adjust drying conditions to optimize quality, for example. Miniaturization of dryers has the advantages of reducing both capital and operating costs. Dryer control using artificial neural nets, fuzzy logic, and model-based control has already become mainstream technology. With deeper understanding of drying processes the operation can be made more reliable and cost-effective. If energy costs skyrocket, drying will be an expensive operation in need of serious R&D!

A number of the author's articles and keynote lectures have listed numerous challenging ideas for academic researchers as well as industrial R&D personnel. For lack of space only a few of the relevant ideas and references are mentioned here for the benefit of the interested reader.

Wang et al. (2004) have summarized the progress in drying technology pertinent to production of nanomaterials using spray drying, supercritical carbon dioxide, freeze drying, etc. They have identified the problems peculiar to the production of nanoparticles. For summaries of R&D needs and future trends the reader may refer to

Table 4 Some R&D problems in selected drying equipment

Dryer type	Nature of R&D problems
Rotary (direct/indirect)	Prediction of particle motion/residence time distribution for poly-disperse solids including effect of cohesion, heat/mass transfer rates; effect of flight design, internal heat exchangers; effect of hold-up; effect of hot air injection into particle bed, noncircular shape of dryer shell, etc.
Fluid bed (direct/indirect)dryers	Effect of particle wetness/poly-dispersity on hydrodynamics, agglomeration, heat/mass transfer rates, etc. Design of internal heat exchangers. Effects of agitation, vibration, etc. Classification of particle types according to fluidization regime including effect of particle wetness/stickiness. Math models. Effect of agitation, vibration, pulsation, acoustic radiation, etc.
Flash dryers	Detailed discrete particle modeling including effects of agglomeration and attrition, effects of geometry, use of pulse combustion exhaust, superheated steam, internal heat exchanger surfaces, variable cross section ducts, hot air injection at various axial locations, CFD models.
Drum dryers	Heat transfer to thin films of suspensions including effects of crystallization, boiling, etc., enhancement of drying rate by radiant heat or jet impingement
Batch dryers	Effects of intermittent heat input using different modes of heat transfer; cyclic pressure swings; variable gas flow; use of heat pumps (including chemical heat pumps), etc
Spray dryers	Effects of various types of atomizers on flow patterns, product properties, agglomeration, size distribution; chamber geometry effects; injection of supplementary air; superheated steam operation; CFD models to investigate novel dryer designs, e.g. horizontal spray dryers; multi-stage horizontal spray/fluid bed dryers
Impingement dryers	Effects of high temperatures (tissue drying); non-circular multiple jets; variable spacing arrays of round or non-circular jets

Mujumdar (2002) and Mujumdar (2004). Nanoparticles can be produced by drying of nano-size droplets, which can be produced utilizing either piezoelectric or thermal actuators. Ashley et al. (1977), Buskrik et al. (1988), and Chen et al. (1999) have discussed the production of fine droplets using piezoelectric actuators. This technology is found in most ink-jet printers except the thermal ink-jet printer manufactured by Hewlett Packard. The HP printer utilizes the concept of explosive boiling where high-frequency heat input generates and collapses very fine bubbles, which cause ejection of nano-size droplets from the printer head. These techniques can be used to produce mono-size micro or nano-scale droplets. When dried they produce nanoparticles. Clearly this technology is new, expensive, and useful only for extremely high-value products such as medical formulations. Professor X.D. Chen at the University of Auckland has succeeded in producing nanoparticles using piezoelectric inkjet technology (Personal communication).

Superheated steam drying is now recognized to become the technology of choice when energy cost and environmental impact as well as product quality are critical criteria. Devahastin et al. (2005) have discussed the use of low-pressure superheated steam for drying heat-sensitive materials such as vegetables, fruits, etc. Such technology is also very significant for drying materials such as coal, which are prone to combustion and explosion if dried in the presence of oxygen. Fluidized bed, flash, or rotary dryers utilizing superheated steam as the drying medium are particularly suitable for industrial drying of materials such as coal, pulp, waste sludges, hog fuel from pulp mills, etc.

Although pulse combustion, like superheated steam drying, is an old technology, its application for drying heat-sensitive materials is relatively recent. This is because, despite numerous advantages, there are severe problems of noise as well as scale-up. Much R&D is needed for this technology to be accepted on a large scale commercially. Wu and Mujumdar (2006) have made a computational fluid dynamic study of drying of droplets in a pulsating high-temperature turbulent flow. Mujumdar and Wu (2004) have examined via simulation the suitability of utilizing a novel pulse combustor chamber based on Nature's design of the Bombardier Beetle. Much experimental as well as modeling research needs to be done on pulse combustion as well as pulse combustion drying. Key problems yet to be analyzed include operation under pressure, use of renewable fuels such as bio-diesel in a pulse combustor and effect of geometric design of the combustion chamber.

More recently heat pump-assisted drying has come of age. A number of commercial suppliers around the world sell such dryers for drying wood, ceramics, fruits and vegetables, spices, marine products, etc. Sun et al. (2005) have carried out experimental and modeling studies of heat pump-assisted drying of vegetable products and shown that a combination of heat sources such as convection and radiation applied intermittently can provide advantages of energy savings as well as improved product quality. Although much of the literature on this topic deals with mechanical heat pumps, Ogura et al. (2003) in a series of papers have shown the potential for the use of chemical heat pumps for batch or semi-continuous drying. Again, much R&D is needed to improve on existing knowledge about mechanical as well as chemical heat pump-assisted drying. It is also possible to enhance the energy efficiency of such systems by coupling storage type heat exchangers to recover heat from the dryer exhaust and using it to preheat the inlet air in the same or next batch. Phase change materials (PCM) provide an attractive means of storing and releasing heat due to their high-density of energy storage in the form of heat of fusion.

Over 25,000 spray dryers are estimated to be in industrial operation around the world. It is difficult to make this operation highly energy-efficient due to the need for large volumes of drying medium (air, superheated steam, or flue gases). Such dryers are notoriously difficult to scale up from laboratory to pilot to full scale. This is because the phenomena involved are highly non-linear. Much of the design and scale up is based on know-how rather than knowledge. However, in recent years successful attempts have been made to utilize computational fluid dynamic software to analyze effects of geometric and flow parameters on spray dryer performance. The reader may wish to consult Huang et al. (2003, 2004) and Huang and Mujumdar (2005) for a series of papers on this topic.

For lack of space no mention is made in this paper of the need for reliable mathematical models for the design, analysis, optimization, and control of industrial dryers. Such models, to be truly useful in practice, must combine transport phenomena with material science and be as predictive as possible. In recent years, the need for stochastic and multi-scale modeling has been recognized but little headway has yet been made in this area. Despite the extensive body of literature on the subject of transport phenomena in porous media, modeling of real drying processes can utilize only a small fraction of this massive effort. A reason for this state of affairs is the fact that drying is product-specific; it may involve chemical as well as physical transformations and such phenomena are rate-dependent and often unpredictable. The mechanisms of moisture and heat transfer in a drying solid can change during the drying process itself. Thus, a significant amount of empiricism appears to be essential even in basic mathematical modeling of drying as well as dryers.

7 Closing remarks

Despite the explosive growth of technical or technological literature on drying, the scientific literature has lagged behind consistently. In recent years, one can discern a new trend; one now can see a number of scientific papers dealing with specific quality attributes of specific products, particularly in the emerging areas of biotechnology and nanotechnology as well as the so-called mature fields of food, agriculture, paper, and wood drying. The list and citation of specific literature is arbitrary and for the purpose of illustration only. Interested readers can find numerous similar examples with an in-depth review of the current literature. It is most important to develop multi-disciplinary research teams for effective R&D in drying.

Techniques such as TRIZ may be useful in utilizing prior experience to develop new solutions to new problems. This technique basically says "borrow ideas from the past and reduce excessive number of features". This Russian-origin technique requires that the user specify the goals rigorously and correctly, focus on key features desired and trim away unnecessary ones, and borrow successful ideas found workable in the past to achieve similar requirements. Any search engine will give interested reader access to many excellent sites that can get one started quickly.

Innovation in drying technologies is continuing over the past two decades although it has not accelerated because of the long life cycles of dryers and relatively unchanged fuel costs over the past decade. No truly disruptive radical drying technologies have emerged, nor are they expected to emerge over the near term. Many innovations are based on existing knowledge viz. use of heat pumps to dry heat sensitive materials using various modes of heat input concurrently. Indeed, the capital and running costs of heat pumps can be reduced by using heat pumps only over the initial drying period beyond which the dehumidified drying air does not enhance the drying rate any longer. Intensification of innovation can occur only when we are able to use a reliable electronic kit to simulate and test new designs without heavy investment of time, manpower, and funds. This implies that a strong base in fundamentals is a pre-requisite to rapid innovation. Finally, some emerging drying technologies that will slowly replace traditional ones are identified, but most of them still require significant R&D effort to be marketable. Interested readers may visit http://www.geocities.com/AS_Mujumdar for information on the latest resources that can assist in their R&D or marketing activity in drying systems.

References

Ashley, C.T., Edds, K.E., Elbert, D.L.: Development and characterization of Ink for an Electrostatic Ink Jet Printer. IBM J. Res. Dev. **21**(1), 69–74 (1977)
Buskrik, W.A., Hackleman, D.E., Hall, S.T., Hanarak, P.H., Low, R.N., Trueba, K.E. Van de Poll, R-R.: Development of high-resolution thermal inkjet printhead. Hewelett-Packard J. 55–61 (1988)
Chen, P.H., Peng, J.Y., Chang, S.L., Wu, T.I., Cheng, C.H.: Pressure response and droplet ejection of a piezoelectric inkjet printhead. Int. J. Mech. Sci. **41**, 235–248 (1999)
Devahastin, S., Suvarnakuta, P., Soponsonnarit, S., Mujumdar, A.S.: A comparative study of low-pressure superheated steam and vacuum dying of heat-sensitive materials. Drying Technol. **23**(3), (2005)
Dodgson, M., Gann, D.M., Salter, A.J.: The intensification of innovation. Int. J. Innov. Manage. **6**, 53–83 (2002)
Foster, R.: Innovation–The Attacker's Advantage. Summit Books New York (1986)
Howard, W.G., Guile, B.R. (eds.): Profiting from Innovation. The Free Press, New York (1992)
Huang, L., Kumar, K., Mujumdar, A.S.: Use of computational fluid dynamics to evaluate alternative spray chamber configurations. Drying Technol. Int. J. **21**(3), 385–412 (2003)
Huang, L., Kumar, K., Mujumdar, A.S.: Simulation of a spray dryer fitted with a rotary disk atomizer using a three-dimensional computational fluid dynamic model. Drying Technol. **22**(10), 1489–1516 (2004)
Huang, L., Mujumdar, A.S.: Development of a new innovative conceptual design for horizontal spray dryer via mathematical modeling. Drying Technol. **23**(6), 1169–1187 (2005)
Kudra, T., Mujumdar, A.S.: Special Drying Techniques and Novel Dryers. Handbook of Industrial Drying, pp. 1087–1149. Marcel Dekker, New York (1995)
Kudra, T., Mujumdar, A.S.: Advanced Drying Technologies. Marcel Dekker, New York (2002)
Mujumdar, A.S.: Superheated Steam Drying– Principles, Practice and Potential for Use of Electricity, Canadian Electrical Association Report 817U671. Montreal, Canada (1990)
Mujumdar, A.S. (ed.): Handbook of Industrial Drying, 3rd edn., pp. 1270. CRC Press, Boca Raton, FL, USA (2006)
Mujumdar, A.S.: Drying research- current state and future trends. Dev. Chem. Eng. Miner. Process. **10**(3/4), 225–246 (2002)
Mujumdar, A.S.: Research and developments in drying: recent trends and future prospects, Drying Technol. **22**(1-2), 1–26 (2004)
Mujumdar, A.S., Wu, Z.H.: Is Bombardier Beetle design suited for a pulse combustor? Topics in Heat and Mass Transfer, International Symposium and Workshop on Industrial Drying. (ISWID2004) Mumbai, India, December, 20–23 (2004)
Ogura, H., Ishida, H., Kage, H., Mujumdar, A.S., Enhancement of energy efficiency of a chemical heat pump-assisted convective dryer. Drying Technol. **21**(2), 279–292 (2003)
Sun, L., Islam, Md. R., Ho, J.C., Mujumdar, A.S.: A diffusion model for drying of a heat sensitive solid under multiple heat input modes. Biores. Technol. **96**(14), 1551–1560 (2005)
Teece, D.J.: Profiting from technological innovation–Implications for integration, collaboration, licensing and public policy. Res Policy **15**, 285–305 (1986)
Wang, B.H., Zhang, W.B., Zheng, W., Mujumdar, A.S., Huang, L.X.: Progress in drying technology for nanomaterials. Drying Technol. **23**(1), 1–18, (2004)
Wu, Z.H., Mujumdar, A.S.: A parametric study of spray drying of a solution in a pulsating high temperature turbulent flow. Drying Technol. **24**(6), 751–761 (2006)

ORIGINAL PAPER

Modern methods of drying nanomaterials

Zdzisław Pakowski

Received: 15 November 2005 / Accepted: 26 March 2006 /
Published online: 30 August 2006
© Springer Science+Business Media B.V. 2006

Abstract This paper presents the types of nanomaterials whose technology requires drying during processing. The methods of drying of three groups of nanomaterials are discussed: nanoparticles, nanolayers (nanofilms), and nanoporous materials. Principally the method of spray drying using ultrasonic nebulizers and electrospraying is presented for the first group. The Langmuir–Blogget technique is presented for the second group. The method of supercritical drying and the solvent replacement technique is described for the third group. Technological aspects of the presented methods are also briefly described.

Keywords Nanoparticles · Nanofilms · Aerogels · Nebulizers

1 Introduction

Nanomaterials are a rapidly growing group of materials, which are composed of elements whose size ranges from 1 to 100 nm. The following main groups of nanomaterials are distinguished:

(1) Nanoparticles.
(2) Nanotubes and nanowires.
(3) Nanodispersions.
(4) Nanostructured surfaces and films.
(5) Nanocrystalline materials.
(6) Nanoporous materials.

Drying can be used in a production cycle of the following groups: nanoparticles, nanostructured surfaces, and films and nanoporous materials. In all cases, the drying process is essential in obtaining the designed and expected properties of the final

---

Z. Pakowski (✉)
Department of Heat and Mass Transfer,
Faculty of Process and Environmental Engineering of Lodz Technical University, Lodz, Poland
e-mail: pakowski@chemeng.p.lodz.pl

product. Moreover, in many cases these properties could not be obtained without a specific drying technique (cf. Pakowski 2004).

When talking about the emerging nanotechnologies a common misinterpretation of the word nano is that, we tend to envisage nanotechnology as being performed on a tiny scale in tabletop factories. On the contrary, some nanomaterials will be produced in tons or more—just take nanoparticles produced as plastic fillers, pigments, varnishes, etc. as an example. Such technology would involve product design from molecular level to the end product, which in turn would require multiscale modeling during the design stage and mastering technological challenges of large scale production—all these new problems are the playground of a modern chemical engineer.

In this work, we try to present some insight into the role that drying plays in the production of such materials. Please bear in mind that, we concentrate on man-made nanomaterials only. Natural nanomaterials (e.g. montmorillonite and other nanoclays, natural zeolites, etc.) are produced by traditional methods: therefore, as the title of this paper implies, only modern methods of drying nanomaterials will be discussed here.

2 Drying at molecular level

Water molecules are approximately 0.35 nm in size. When talking about nanomaterials water certainly cannot be considered a continuum. However, even in nanotubes 50 nm across (equivalent to 150 water molecules) menisci of water are observed (Gogotsi et al. 2001) i.e. interaction forces between water molecules and between water molecules and solid result in a continuous surface of water. These liquid menisci are responsible for capillary pressure and drying stresses in nanoporous solids. They are also responsible for water movement in nanosuspensions (Dufresne et al. 2003). It was observed that in nanosuspensions of 6, 11, and 26 nm particles of up to 30% mass concentration in water, drying is never diffusive in nature and menisci never penetrate the solid. Instead, water moves towards the surface by Darcy flow and the falling drying rate period is a result of increasing resistance to flow caused by compaction.

Drying a thin layer of suspension of nanoparticles in liquid on a molecular level involves a play of interaction forces between molecules of gas, molecules of liquid, and nanoparticles, which are only several times larger. The surface tends to assume an equilibrium state in which the overall surface energy is the lowest. When molecules of water are removed from the surface one by one during drying, new surface configurations emerge. Finally nanoparticles form drying-induced structures by this self-assembly process (Sear et al. 1999, Rabani et al. 2003). The process can easily be modeled by Monte–Carlo simulation and the surface structures obtained closely resemble the structures obtained by drying (Rabani et al. 2003). The self-assembly processes can exist at all scales (Whitesides and Grzybowski 2002), but it can be very helpful at nanometric scale in obtaining custom-designed structures (cf. next section).

Understanding of drying at the molecular level can be very helpful in drying at the nanometric scale and can be used to obtain the designed product properties.

3 Nanoparticles

3.1 Applications

Nanoparticles are probably the most promising group of nanomaterials and have immediate applications. The estimated total world market for nanoparticles (Rittner 2002) was $492.5 million in the year 2000 and it is estimated to reach $900.1 in 2005. Possible applications are endless; Table 1 presents the most important predicted applications.

Popular high-tonnage applications as plastic or man-made fiber fillers are not even mentioned. Already today, numerous products are advertised as containing nanoparticles, among them, e.g. refrigerators lined with plastic coating containing silver nanoparticles having aseptic properties are becoming popular.

3.2 Nanoparticle technology

Nanoparticle production uses basically two methods: gas–particle conversion and liquid–particle conversion (Pakowski et al. 2005). The first group forms the so-called dry route, where particles are obtained from vapor by condensation, from gases by reaction, or from plasma by deposition. A clear advantage of the dry route is that it allows the production of very small particles (1–20 nm); however, not all materials can be processed, especially temperature sensitive ones.

The second group is the so-called wet-route, where spray drying occupies the most prominent place (Okuyama and Lenggoro 2003). This group also includes supercritical expansion in a nozzle or supercritical expansion into liquid nitrogen, among other more exotic methods. The advantages of the wet route are that it allows processing of temperature sensitive, multicomponent and bioactive materials, but the particles obtained are larger and they often agglomerate.

Table 1 Predicted applications of nanoparticles

Electronic, optoelectronic, magnetic applications	Biomedical, pharmaceutical, cosmetic applications	Energy, catalytic, surface coating applications
Chemical-mechanical polishing	Antimicrobials	Automotive catalysts
Electroconductive coatings	Biodetection and labeling	Ceramic membranes
Magnetic fluid seals	Biomagnetic separations	Fuel cells
Magnetic recording media	Drug delivery	Photocatalysts
Multilayer ceramic capacitors	MRI contrast agents	Propellants
Optical fibers	Orthopedics	Scratch-resistant coatings
Phosphors	Sunscreens	Structural ceramics
Quantum optical devices		Thermal spray coatings
Solar cells		Dirt-free windows
		Antibacterial surfaces

3.3 Nanoparticles by spray drying

Spray drying is a well-mastered technology, but the particles obtained in conventional spray drying are in the micrometric range. In order to reduce their size to nanometric range, two basic steps are necessary:

(1) Reduction of the initial spray size—this is available using spraying techniques that allow the generation of much smaller droplets—ultrasonic nebulization produces droplets ca. 1.5 µm in size while electrostatic spraying (Lenggoro et al. 2000) allows the production of even smaller droplets.
(2) Low initial concentration—when droplets initially contain mostly solvent they shrink during drying. To obtain 100 nm particles from 5 μm droplets of solution in water the initial concentration must be as low as 0.000008% (Iskandar et al. 2003).

It is obvious that particles produced from solutions may have amorphous and porous structures, which is often unwanted. In order to obtain spherical particles, it is advised to refer to sol–gel technology, where particles are produced from a sol using a suitable precursor. After condensation of the precursor a colloidal suspension of spherical particles is formed, which can be spray dried after thinning with solvent to a necessary initial concentration.

When spray droplets contain nanoparticles the interaction forces between them and solvent molecules may lead to self-assembly in the same way as on a flat surface. This idea was exploited in order to produce structured empty microparticles composed of nanoparticles and resembling a golf ball (Iskandar et al. 2001, 2003). The particles were produced by spray drying a suspension of ca. 20 nm silica particles mixed with 79 nm latex particles. The process of self-assembly led to the formation of a regular hexagonal structure where the core of each hexagon was occupied by the latex particle. The latex particles were then sublimated at high temperature during a single pass through the spray dryer-calciner, leaving pores behind. It was also found that by controlling of the process conditions the resulting particles can be spherical or toroidal in shape (Iskandar et al. 2003). This opens numerous possibilities in custom design of the end product.

It has to be added that collection of the product obtained can only be achieved in electrostatic precipitators, since other methods of separating such fine aerosols are inadequate (filtration and inertial methods are ineffective). Care should be taken when working with nanoparticles since inhaled aerosols may end up in the lungs and easily penetrate the vacuole. It is commonly accepted that at this scale of dispersion nanoparticles can have adverse effects, even if they are composed of nontoxic material.

It is interesting from the chemical engineer's point of view to know how the spray dryer for nanometric particles can be designed. It is obvious that typical correlations for the heat or mass transfer coefficients of Ranz and Marshall will not hold because the Reynolds number is too small. Even the use of calculus is questionable, since it involves the idea of a continuum, which does not hold at the nanoscale. This calls for methods of molecular dynamics and Monte–Carlo simulations and, when applied to design, it certainly calls for multiscale modeling, which is now emerging as a design tool. Spray drying of nanoparticles will certainly be an elegant subject to test this approach.

3.4 Other methods of drying nanoparticles

Bulk drying methods like oven drying are sometimes used for drying nanoparticles at the laboratory scale. For that purpose all methods used for drying gels (cf. Sect. 5) can also be used. In order to avoid agglomeration, all ions and water have to be removed prior to drying. Ions are removed by washing with de-ionized water and water is removed by organic solvents with possibly low surface tension and functional groups that possibly replace hydroxyl groups on the surface of the nanoparticles, which are partly responsible for agglomeration. Since these solvents do not all mix with water, a sequential solvent replacement is usually performed. Supercritical drying can also be used for nanoparticles (Wang et al. 2005).

4 Nanolayers and films

Nanolayers and films are one of numerous applications of nanoparticles. They can be produced, e.g. by plasma sputtering, but the most popular method is a wet route using the Langmuir–Blogget technique. In this method, a suspension of nanoparticles in a carrier liquid is poured over a surface of water when it forms a "monomolecular" layer. The layer is transferred to a solid surface by dipping and the solvent is removed by drying.

Possible applications include sensors, especially for biomedical applications, tribologic applications, etc. Drying of such layers is decisive for the quality of the final product. Multilayer nanofilms have been produced by the self-assembly technique (Borato et al. 1997), in which protein molecules can be built into films of conducting polymers. It allows one to build nanofilms on a quartz substrate by repeated dipping in polycation and polyanion solutions, separated by washing and drying. A multilayer film of lysozome and polystyrene sulfonate can be made in this way. The drying stage directly affects the amount and quality of adsorbed material. Dehydration of proteins results in significant conformational changes that are only partly reversible (Prestrelsky et al. 1993). Rapid drying leaves the lysozome layer "shrunk" and discontinuities in the film may appear, leaving space for other molecules to absorb. This gives possibilities for controlling the film structure.

Drying of a film containing nanoparticles on a liquid surface is a stage where advantage can be taken of self-assembly to obtain a regular film structure. Evers et al. (2002) produced perfect films of nanoparticles on a layer of perfluorinated oil over a surface of water where it forms a monoparticle layer. The film was then transferred on a silica substrate and the oil removed by drying. This technique produced perfect monoparticle layers.

5 Nanoporous solids

5.1 Classification of nanoporous solids

The term 'nanoporous solid' is a neologism. The IUPAC classification uses the nomenclature shown in Table 2. Actually nanoporous solids are either microporous or mesoporous according to the IUPAC classification.

Terminology	Pore diameter d
Micropores	$d < 2$ nm
Mesopores	$2 < d < 50$ nm
Macropores	50 nm $< d$

Table 2 Pore classification according to IUPAC

The microporous solid class is almost entirely reserved for zeolites. Zeolites are, however, produced in bulk and are strong enough so that their drying is not a challenging technical problem. A much more interesting class of materials is gels, which are mesoporous. The density of the solid skeleton of gels can vary from a few to a few hundred kg/m^3. Such a skeleton is unable to bear the extreme drying stress and the gels shrink considerably during drying. The only possibility of retaining the original gel structure during drying is by elimination of surface tension, which is possible in supercritical drying.

From the point of view of the pore structure, mesoporous solids can be classified into amorphous and fractal ones. Amorphous solids are typically obtained by sputtering while fractal mesoporous solids are produced by gelation. Dry gels, whose internal structure is retained, are called aerogels.

5.2 Methods of drying

As explained in Sect. 2, drying of nanoporous solids produces an immense drying stress. There are several ways to reduce this drying stress and according to the degree of this reduction the gels obtained are called:

(1) Xerogels, obtained by traditional convective or microwave drying under normal or reduced pressure—their structure is not preserved and shrinkage is enormous.
(2) Cryogels, obtained by freeze drying—their mesopores are destroyed and replaced by macropores resulting from sublimation of ice crystals.
(3) Aerogels, obtained by supercritical drying or sequential solvent replacement drying (cf. Wang et al. 2005) and similar methods—their mesoporous structure is preserved.

Supercritical drying is in principle a perfect method of drying gels. When the pressure and temperature of the sample are raised above the critical point of moisture in an autoclave, surface tension disappears, and solvent can be evacuated slowly without affecting the structure. However, supercritical drying is expensive and ways are sought to eliminate it.

5.3 Supercritical drying

The principle of supercritical drying is the elimination of surface tension by passing the critical point of moisture. This can be achieved in basically two ways:

(1) By supercritical transition of the original solvent—this is not applicable to hydrogels, since water has extreme critical parameters, but may be applied to alcohols, which are often used in gel synthesis. The gels obtained as inorganic oxides (SiO_2, V_2O_5, Al_2O_3) after drying have partly esterified hydroxyl groups, which results in their hydrophobicity.

(2) By supercritical transition of CO_2 or other low-critical parameter solvents—CO_2 is placed into the gel by solvent replacement (washing), which may be either supercritical or subcritical. CO_2 has low critical parameters and therefore the process is applicable even to temperature sensitive gels (organic gels). However, CO_2 produced gels retain their hydroxyl groups and hydrophobization by post-processing is necessary (Pakowski and Maciszewska 2003). In this method, the dry aerogel is placed in Tetramethylsilane (TMS) vapors for several hours and hydroxyl groups are replaced by TMS groups.

Ways are sought to eliminate supercritical drying, since it is too expensive to perform on an industrial scale. In order to eliminate damage caused by shrinkage, ways of reinforcing the structure are investigated in the first place. Three groups of methods emerged as follows:

(1) Maturing the gel in the solution of its precursor until all chemical bonds are complete and the structure building is terminated (Einarsrud 1998).
(2) Using dendrimeric precursors, which have a much more regular structure (Fox et al. 2002).
(3) Using drying control chemical additives (DCCA) to narrow pore size distribution and slow down drying (Pakowski and Bartczak 1997).

Depending on the application, when the quality of the aerogel can be compromised, the methods of convective drying are acceptable. This is usually so in the production of catalysts but is not acceptable in monolithic aerogels.

In some instances, convective drying of gels under controlled drying conditions, so that crack formation is eliminated but shrinking allowed, leads to microporous glasses obtained in the so called cold way by sol–gel technology. For such a process, the drying rate must be lowered considerably by evaporation in a solvent rich atmosphere, which is only very slowly vented to atmosphere. During, such a process the mesopores shrink into micropores leaving porous glass behind (e.g. Vycor®).

Aerogels can be produced as thin layers on electronic substrates but their direct large scale application can be filtration of airborne nanoparticles (Pakowski et al. 2001). Such filters are extremely efficient (they fall into the ULPA category); however, they are still rather brittle and too delicate for practical application. Elastic aerogels, which would solve the problem, have not yet been reported. One company advertises silica aerogel thermal insulation blankets and garments; this however, is probably a blister or pocket type textile with pockets filled with aerogel powder.

6 Technical impact

It has to be realized that almost all the literature available on the subject treats laboratory scale experiments and not a commercialized technology. Being a hot subject, the technology of nanomaterials is not revealed by their manufacturers. But nanomaterials are certainly already manufactured on a large scale, as we can judge from the products that appear on the market. It seems that three major groups of methods have been commercialized as follows:

(1) Spray drying—this method is reported in company literature by certain nanoparticle manufacturers. On the industrial scale it was necessary to solve the problem of increasing throughputs and reducing agglomeration of particles.

(2) Freeze drying—this is applicable both to nanoparticles, which are sprayed directly into liquid N_2 and then freeze dried, or to gels, with their internal structure altered by ice crystal formation. Applicable to thermolabile materials, including pharmaceuticals.
(3) Supercritical drying—already used by a few companies to produce commercial silica aerogels in the form of aerogel windows or aerogel thermal insulation. Since, aerogels are attractive high-tonnage materials their share in the market will increase and methods of lowering the overall costs will be sought, e.g. by replacing with convective drying of strengthened gels.

Scaling up of drying processes described above, for reasons described earlier, is difficult and requires a lot of intermediate scale experiments. Therefore, the new technologies emerge slowly and it usually takes several years from the first announcement to the matured technology.

7 Summary

Nanotechnology is here to stay. Drying is a process that is closely related to many of the technologies of nanomaterials and in many instances it determines the final properties of the product. The product is the effect of numerous forces and processes acting on a molecular level that need to be better understood in order to understand the process of drying nanomaterials. Methods of molecular dynamics and multiscale modeling have to be used in order to scale up the process. Numerous methods for drying nanomaterials already exist, which sometimes differ significantly from the classical method of drying. This is why the subject of drying nanomaterials requires more attention and the scientific incentives to study it the nearest future will be certainly accompanied by increasing funding from international agencies, governments and industry.

References

Borato, C.E., Herrmann, P.S.P., Colnago, L.A., Oliveira, O.N. Jr. Mattoso, L.H.C.: Using the self-assembly technique for the fabrication of ultra-thin films of a protein. Braz. J. Chem. Eng. **14**, 4 (1997)

Dufresne, E.R., Corwin, E.J., Greenblatt, N.A., Ashmore, J., Wang, D.Y., Dinsmore, A.D., Cheng, J.X., Xie, X.S., Hutchinson, J.W., Weitz, D.A.: Flow and fracture in drying nanoparticle suspensions. Phys. Rev. Lett. **91**(22), 224501-1–224501-4 (2003)

Einarsrud, A.-A.: Light gels by conventional drying. J. Non-Cryst. Solids **225**, 1–7 (1998)

Evers, M., Schöpe, H.J., Palberg, T., Dingenouts, N., Ballauff, M.: Residual order in amophous dry films of polymer lattices: indications of an influence of particle interaction, J. Non-Cryst. Mater. **307–310**, 579–583 (2002)

Fox, G.A., Baumann, T.F., Hope-Weeks, I.J., Vance, A.L.: Chemistry and processing of nanostructured materials. DOE report UCRL-ID-146820 (2002)

Gogotsi, Y., Libera, J.A., Yazicioglu, A.G., Megaridis, C.M.: In-situ Multiphase Fluid Experiments in Hydrothermal Carbon Nanotubes. Appli. Phys. Lett. **79**, 1021–1023 (2001)

Iskandar, F., Gradon, L., Okuyama, K.: Control of morphology of nanostructured particles prepared by the spray drying of nanoparticle sol. J Colloid Interf Sci **265**, 296–303 (2003)

Iskandar, F., Mikrajuddin, A., Okuyama, K.: In situ production of spherical silica particles containing self-organized mesopores. Nano Lett. **1**(5), 231–234 (2001)

Lenggoro, I. W., Okuyama, K., Fernandez de la Mora, J., Tohge, N.: Preparation of ZnS nanoparticles by electrospray pyrolysis. J. Aerosol Sci. **31**, 121–136 (2000)

Okuyama, K., Lenggoro, I.: Preparation of nanoparticles via spray route. Chem. Eng. Sci. **58**, 537–547 (2003)

Pakowski, Z., Bartczak, Z.: Modeling of multicomponent drying of a shrinking gel cylinder containing DCCA. Dry Technol. **15**(2), 555–573 (1997)

Pakowski, Z., Nowacka, U. Abo Zebida, O., Głębowski, M.: The investigation of the efficiency of aerogel filters in the nanometric range (in Polish). Inż. Chem. Proc. **22**(3D), 1085–1090 (2001)

Pakowski, Z., Maciszewska, K.: The evaluation of methods of hydrophobization of fibrous filters with deposited silica aerogel layer (in Polish). Przem. Chem. **82**(8–9), 1243–1245 (2003)

Pakowski, Z.: Drying of Nanoporous and Nanostructured Materials. IDS'2004. São Paulo, Brazil (2004)

Pakowski, Z., Czapnik, M., Piątkowski, M., Zbiciński, I.: Production of Nanoparticles by Spray Drying (in Polish). XI PDS, Poznań (2005)

Prestrelski, S.J., Tedeschi, N., Arakawa, T., Carpenter, J.F.: Dehydration-induced conformational transitions in proteins and their inhibition by stabilizers. Biophys. J. **65**, 661–671 (1993)

Rabani, E., Reichman, D.R., Geissler, P.L., Brus, L.E.: Drying-mediated self-assembly of nanoparticles. Nature **426**, 271–274 (2003)

Rittner, M.N.: Market analysis of nanostructured materials. Am. Ceramic Soc. Bull. **81**, 3 (2002)

Sear, R.P., Chung, S.-W., Markovich, G., Gelbart, W.M., Heath, J.R.: Spontaneous patterning of quantum dots at the air-water interface, Phys. Rev. E. **59**(6), R6255–6258 (1999)

Wang, B., Zhang, W., Zhang, W., Mujumdar, A.S., Huang, L.: Progress in drying technology for nanomaterials. Dry. Technol. (**23**) 1–2, 7–32 (2005)

Whitesides, G.M., Grzybowski, B.: Self-assembly at all scales. Science **295**, 2418–2421 (2002)

ORIGINAL PAPER

Internal heating effect and enhancement of drying of ceramics by microwave heating with dynamic control

Yoshinori Itaya · Shigeru Uchiyama · Shigekatsu Mori

Received: 5 November 2005 / Accepted: 26 March 2006 /
Published online: 30 August 2006
© Springer Science+Business Media B.V. 2006

Abstract The effectiveness of internal heating for enhancing the drying of molded ceramics is evaluated by both modeling and experiments. In the theoretical analysis, three dimensional drying-induced strain–stress are modeled, and the numerical solutions show that the internal heating generates lower internal stress than continuous convective heating or intermittent convective heating. Microwave drying is examined experimentally to study the effect of internal heating on the drying behavior of a wet sample of a kaolin slab. The drying behavior is compared among three modes: microwave heating, hot air heating and radiation heating. The transient behavior of temperatures in microwave drying is quite different from conventional drying by external heating. In particular, the temperature of the slab drops once in the progress of drying. This phenomenon cannot be predicted adequately by a simple model of one-dimensional heat conduction and moisture diffusion accompanied with an internal heat generation rate given as a linear function of the moisture content. It should be noted that the temperature behavior takes place due to the combined interactions with internal evaporation of moisture by rise in internal vapor pressure and shift of impedance or interference in the applicator. Microwave heating with a constant power above 100 W results in sample breakage due to the internal vapor pressure. However, if the power is dynamically controlled so as to maintain the temperature less than the boiling point of water, the drying succeeds without any crack generation until completion with a significantly faster drying rate than drying in convective heating or in the oven.

Y. Itaya (✉) · S. Uchiyama
Department of Chemical Engineering
Nagoya University
Nagoya 464-8603, Japan
e-mail: yitaya@nuce.nagoya-u.ac.jp

S. Mori
Center for Cooperative Research in Advanced Science & Technology
Nagoya University
Nagoya 464-8603, Japan

Keywords Microwave drying · Internal heating · Ceramics · Molding · Crack formation · Heat and mass transfer · Dynamic control · Drying enhancement

Nomenclature

Bi	Biot number [−]
C_0	Initial moisture content defined by water weight per unit volume of layer [kg/m^3]
$C*$	Dimensionless moisture content normalized by initial moisture content [−]
c_p	Specific heat of layer [J/(kg·K)]
c_{pw}	Specific heat of water [J/(kg·K)]
f_v	Ratio of overall mass transfer coefficient between interface and air to that between surface and air [−]
H_a	Humidity of air [kg-H$_2$O/kg-dry air]
H_s	Saturated humidity at surface temperature of layer [kg-H$_2$O/kg-dry air]
Le	Lewis number [−]
$Q*$	Dimensionless internal heat generation rate [−]
Q_0*	Dimensionless internal heat generation rate of sample with initial moisture content [−]
r_w	Latent heat of water evaporation [J/kg]
T	Local temperature [K]
T_0	Initial temperature [K]
T_a	Air temperature [K]
$\bar{T}$	Average temperature of sample [K]
$T*$	Dimensionless temperature normalized by initial temperature [−]
T_a*	Air temperature normalized by initial temperature [−]
V	Volume of a slab [m^3]
w_0	Initial moisture content of a slab with dry basis [kg-H$_2$O/kg-dry solid]
$\bar{w}$	Average moisture content of a slab with dry basis [kg-H$_2$O/kg-dry solid]
$x*$	Coordinate in depth direction (dimensionless depth from surface normalized by thickness of layer) [−]
x_e*	Dimensionless depth of interface from surface [−]
θ	Dimensionless time defined as Fourier number [−]
ρ	Density of layer [kg/m^3]
ϕ_1	Correction factor defined by ratio of thermal diffusivity in dry layer to wet layer [−]
ϕ_2	Correction factor defined by ratio of thermal conductivity in dry layer to wet layer [−]
φ	Proportional constant in Eq. (10) [−]

1 Introduction

Drying of ceramics involves handling molded materials, and it is essentially different from drying granules and powders. Hence, the drying processes generally require a long time as internal moisture movement in bulk controls the drying rate and the designed configuration should be maintained precisely without any failure or irregular deformation. To advance the drying rate, drying time and quality of dry products, internal moisture transfer must be enhanced by controlling to match the heat and mass transfer in a boundary film layer on the materials (Gong et al., 1998; Itaya et al. 1997). However, it is not usually very easy to directly enhance the internal moisture transfer. Itaya et al. (1999, 2001) studied the influence of intermittent heating on the formation of drying-induced strain–stress in molded slabs by parametric analyses. However, the effectiveness of intermittent heating was not sufficient to reduce the internal stress. If molded ceramics are internally heated, then internal vapor pressure would rise, which may result in enhancement of moisture movement. Microwave is a promising candidate for internal heating. Microwave dryers are developed conventionally for powders, granules, blocks, sheets, foods etc. The effect of microwave on the drying of porous materials was studied experimentally by Araszkiewicz et al. (2004). Perre and Turner (1996) studied microwave drying combined with steam or infrared heating to examine its effectiveness. Islam et al. (2003) modeled the drying of shrinking slabs by combined heating by convection, conduction, radiation and microwave. Theoretical analyses solving Maxwell equations were carried out to predict an electric field in microwave dryers by Turner and Jolly (1991), Turner (1994), Liu et al. (1994), and Lehne et al. (1999). Kowalski and Rybicki (2004) and Kowalski et al. (2004) studied the theoretical modeling of moisture transfer and drying-induced stress in saturated porous materials heated by convection and microwave. Research has also been done by Itaya et al. (2003) on an attempt employing random reflection in a fluidized bed of electrically conductive beads to improve uniformity of the electric field intensity distribution of microwave in applicators.

This paper presents the modeling of the behavior of drying-induced strain–stress of ceramics molded in a slab subjected to several heating modes as well as the experimental results of microwave drying, which involve partly reviewing the authors' past works. The influence of internal heating on the drying characteristic is preliminarily specified based on the numerical analysis by the model. The enhancement effect of ceramic drying and the behavior of drying characteristics and failure generation are examined by conducting microwave heating controlled dynamically as a desirable method of internal heating.

2 Analysis of effect of internal heating

Parametric analyses of the strain–stress relationship with heat and moisture transfer in the drying process were performed. The transient three-dimensional problem of strain–stress as well as heat and moisture transfer in a slab of clay was solved simultaneously by the finite element method. A similar model to that of Itaya et al. (1997, 1999) was introduced. The following assumptions were used to simplify the model: (1) the analysis is limited to preheating and constant drying rate periods since major shrinkage of clay takes place in these periods, (2) the Dufour effect is neglected, (3) the initial temperature and moisture content profiles are uniform, (4) water evaporates only on the surface, (5) heat and mass diffusivities are isotropic, (6) the mass transfer equation is expressed by an overall diffusion equation, (7) the sample is a linear viscoelastic material, (8) drying-induced shrinkage is given by a function of the local moisture content only. The governing equations of the heat and mass transfer are deduced in dimensionless forms. The more detailed explanation and the formulae of the model are summarized here.

Fig. 1 Transient behavior of average moisture content and temperature

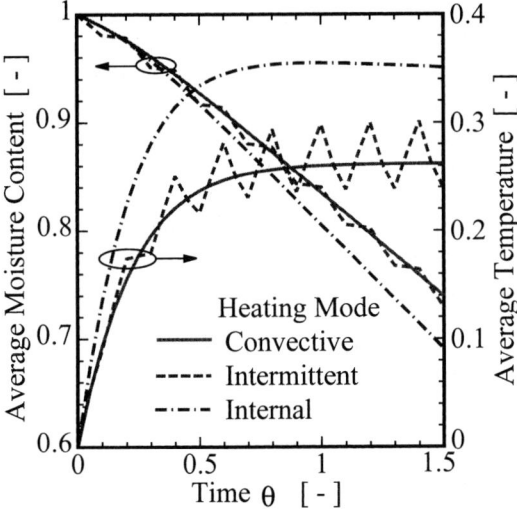

Fig. 2 Maximum tensile stress against average moisture content

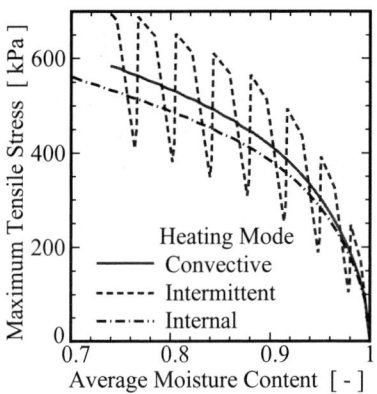

The theoretical analyses were performed for three modes: (1) continuous hot air heating, (2) intermittent hot air heating and (3) internal heating. The intermittent heating mode was modeled so that a slab is heated alternately with a high and low Biot number (0.2 and 0.02, respectively) in every $\Delta\theta = 0.1$ interval dimensionless time (Fourier number). This model assumes that the convective heat transfer coefficient between a slab and hot air varies due to alternately changing the flow rate of hot air. A constant Biot number at 0.11 was given for the continuous heating mode. The internal heating rate given into the body was assumed to be equal to the continuous heating mode at the beginning of drying. The properties and the parameters necessary for the analysis were same as those in the past papers by the present authors (Itaya et al. 1997, 1999).

The transient behaviors of average moisture content and temperature in a slab analyzed numerically are shown in Fig. 1. The moisture content and temperature are given here in dimensionless forms defined by $\overline{w}/w_0$ and $(\overline{T} - T_0)/(T_a - T_0)$, respectively. The average temperature is determined by the integration average over the body of the slab as: $\overline{T} = \int_V T \, dV/V$. The moisture content and temperature in the intermittent heating mode fluctuates over and under the results in the continuous convective heating mode. In the internal heating mode, both the drying rate and the temperature rise significantly compared with the other modes. The reason can be explained by the fact that the net heating rate falls gradually due to reduction of the temperature difference between hot air and the surface with elevation of the surface temperature in the convective heating while the internal heating is adopted with a constant rate through the drying period in this model. The maximum tensile stress is plotted against the moisture content in Fig. 2 to compare the behavior among the heating modes. Intermittent heating contributes to a remarkable fall of the stress in the low heating period, but the stress results in loading much strongly on the slab in every high heating period. The line is not smooth as each data point is connected by a straight line, but a rise in the tensile stress is rather large right after turning to the high heating period than just before turning to the low heating period, since the gradient of the moisture profile is significant on the surface of the slab. In the internal heating mode, the maximum stress is smaller than the convective heating, particularly in the later period of drying. In general, as convective drying proceeds rapidly at the corners of the slab that become the highest temperature in the body, the maximum stress is formed around these portions. However, when the body is heated internally, the temperature becomes the lowest at the corners, and the drying rate is minimized there as well. This fact is considered as a reason why the internal heating forms lower stress than external heating. The present analytical model assumes no influence of temperature on the apparatus diffusivity of moisture in the slab. However, as vapor pressure actually increases exponentially with a rise in temperature, the moisture movement in the body will be enhanced by the internal pressure diffusion of vapor. Hence, it is implied that the internal heating mode may be the most effective one for drying molded materials among those heating modes under a limited simulation in the present work if the tensile stress can be related to the cause of crack generation.

3 Experiment

Figure 3 shows an outline of the experimental setup for examining the behavior of microwave drying of a slab made of clay. The applicator is made of a steel duct 100×100 mm in cross section and 500 mm in length. The test section is located in the center of the duct. A plate $\phi 50$ mm as a sample holder is connected to an electronic balance to measure the time behavior of sample weight during drying. Microwave generated in a magnetron (Tokyo Electronics Inc., IMG-2501S) is irradiated to the sample through the waveguide joint on the test section from above. The frequency is 2,450 MHz and the power can be varied dynamically in the range 100–1000 W. An isolator, a stab-tuner and a power monitor are installed for tuning microwave between the magnetron and the applicator. Steel wire nets are set at both sides of the test section in the duct to seal the microwave from leakage. Convective heating can take place simultaneously by flowing hot air in the duct of the applicator. The aperture at the joint between the duct and the waveguide is covered with a glass plate so as not to disturb hot air flow in the test section.

Drying by convective heating is performed without microwave irradiation using the same dryer as Fig. 3. The drying behavior due to microwave heating is also compared with the data due to radiative heating in an oven type of dryer (Yamato, Drying Oven DX31).

Fig. 3 Outline of applicator and slab sample for microwave drying experiment. (**a**) Experimental setup and (**b**) Slab sample

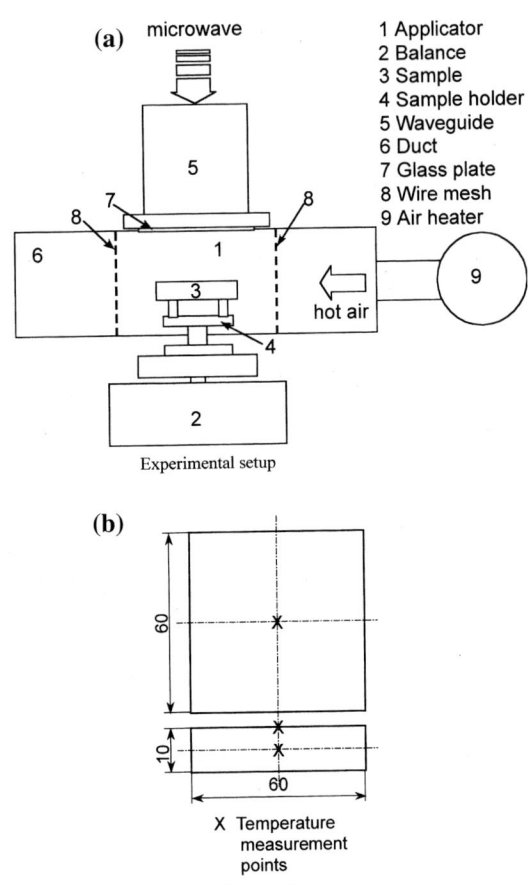

The samples are prepared by molding kaolin into a slab as follows: dry kaolin powder (50 g) is mixed with 20 g water and the wet material is pressed in a mold made of aluminum with pressure 200 kPa. The slab size is 60 × 60 × 10 mm. The median kaolin particle size is 0.75 μm.

The sample is placed horizontally keeping a clearance with small spacer on the holder in the test section so that a surface of the slab faces the incident microwave and both sides are subjected to a parallel air flow. The drying rate is determined from the time behavior of the sample's weight measured by the electronic balance. The transient temperatures are measured by luminescence types of optical fiber thermometer (Adachi Instrument, FX-8500) placed at the center core and the surface of the sample, as seen in Fig. 3(b).

4 Drying characteristic of a slab heated by hot air or radiation

Figures 4 and 5 show the drying behaviors of a kaolin slab heated by hot air in the microwave dryer without irradiation and by radiation in the oven dryer. The moisture content of the sample is plotted against the drying time in the hot air temperature of

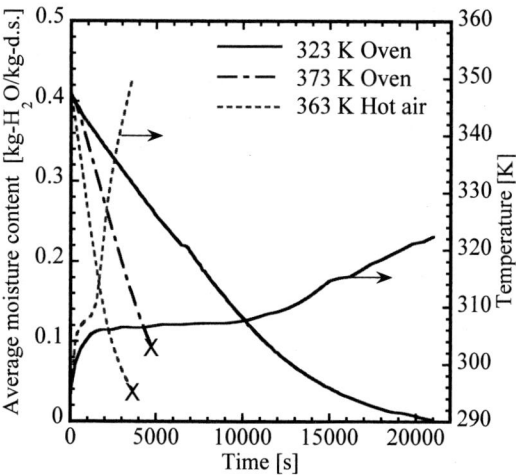

Fig. 4 Drying behavior for hot air heating and for radiative heating in oven dryer

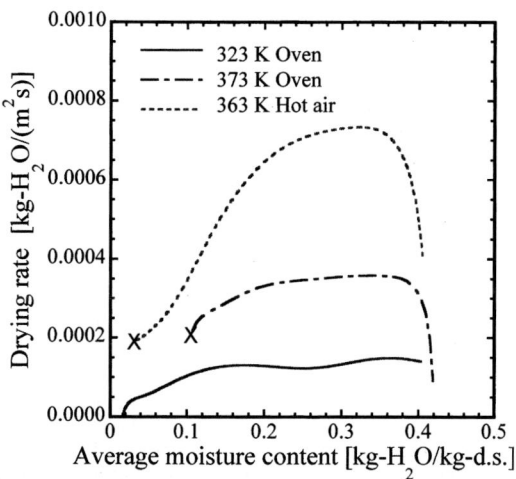

Fig. 5 Drying characteristic curves for hot air heating and for radiative heating in oven dryer

363 K and in the oven temperatures of 323 and 373 K in Fig. 4. The air flow rate is 0.643 m/s for hot air heating. The temperatures are also shown at the position of the center core inside the sample for the cases of the oven temperature 323 K and hot air heating of 363 K. An apparent constant rate period is seen in all runs. This period is also recognized in the drying characteristic curves drawn as the moisture content vs. the drying rate in Fig. 5. The drying in the oven at temperature 373 K proceeds remarkably more slowly than convective heating when compared with the hot air temperature of 363 K. This difference is described by the heating rate to the slab. Indeed, the apparent heat transfer coefficient between the slab and the atmosphere in the constant rate period is 14.6 W/(m²·K) for the oven at 373 K where the atmosphere is stagnant or natural convection while it is 35.7 W/(m²·K) for forced convective heating by hot air at 363 K. Cracks on the sample were generated in hot air heating at 363 K and in the oven maintained at 373 K at the drying times marked by "X". Drying completes successfully without crack formation in the oven at 323 K, but the drying takes almost 6 hours.

Fig. 6 Transient moisture content and temperature of kaolin slab during microwave drying

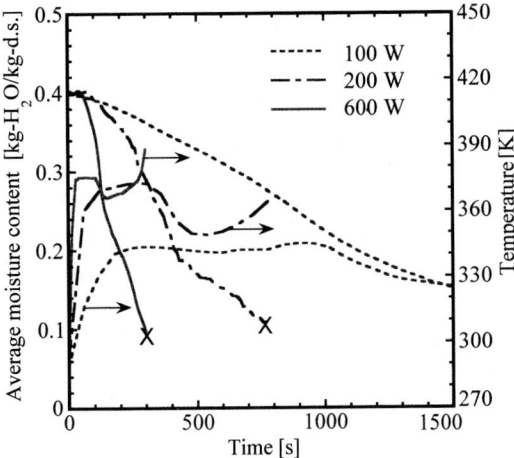

Fig. 7 Drying characteristic curves of kaolin slab during microwave drying

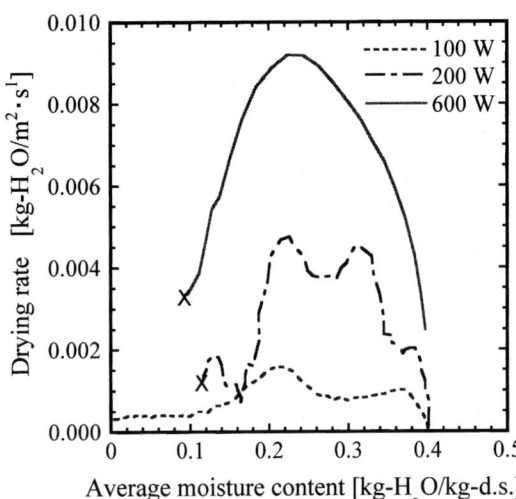

5 Microwave drying with constant power

Figures 6 and 7 show the transient behavior of the moisture content and the internal core temperature of a kaolin slab during microwave drying with a constant power from 100 to 600 W. Air flows in the applicator at atmospheric temperature. It is noticed that the behavior is significantly different from drying by hot air heating or in the oven. In particular, the temperatures increase rapidly at first until being held at a certain value depending on the power at the time. But the temperatures fall once and subsequently start to rise again. The drying rate shows a complex behavior similar to the temperature, as seen from the drying characteristic curves in Fig. 7. These phenomena are considered to be a specific feature due to microwave heating. In an earlier stage, the sample contains sufficient moisture, so microwave energy is absorbed well in the body. When the drying proceeds and the moisture content is reduced, the absorption of microwave radiation in the sample falls while the drying rate may accelerate due to

Fig. 8 Photographs of crack generated on samples during drying. (**a**) drying by microwave heating and (**b**) drying by hot air heating

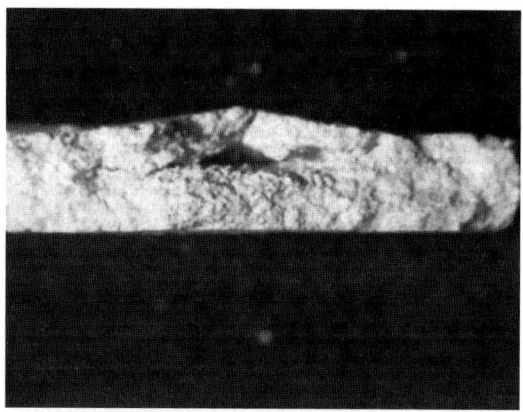
(a) drying by microwave heating

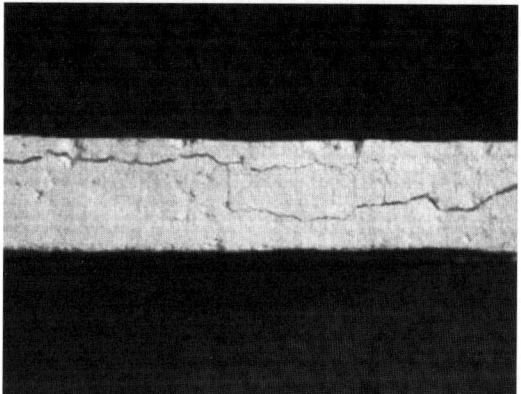
(b) drying by hot air heating

the driving force of a rise in the vapor pressure. The impedance at the test section in the applicator may also shift somewhat out of adjustment with the incident microwave because of the change of the moisture content distribution. These reasons will result in a fall in the temperatures once during the process of drying. Reduction of the drying rate and re-adjustment of impedance will be a reason why the temperatures rise again when the drying proceeds further. Any kaolin sample heated by power of over 200 W was broken when the internal temperature achieved almost the boiling point or 373 K. No crack generates on the sample for 100 W, at which power the temperature is never raised to 370 K. It still takes a long time for completion of drying, though the drying time improved rather than the external heating. Crack formation is caused by expansion due to elevation of internal vapor pressure in microwave drying while a crack forms by tensile stress on the shrinking surface in hot air or radiative drying, as observed from the photographs in Fig. 8.

6 Transient temperature during internal heating drying

It was observed in the above section that a unique temperature behavior appears in microwave drying of slabs. The transient behavior is studied further to discuss the

Fig. 9 Outline of one-dimensional model

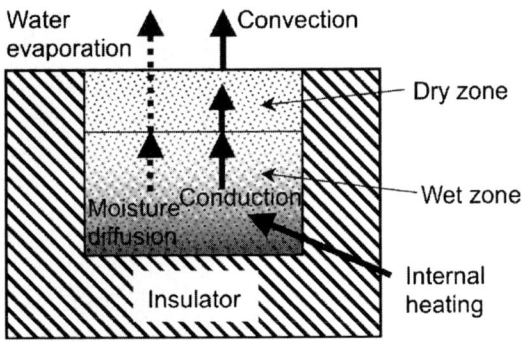

mechanism in terms of a model. Ignoring drying-induced shrinkage in this work, one-dimensional heat conduction and moisture diffusion equations are solved numerically in the dimensionless forms to simplify the model. Indeed, the properties of clay are dependent on moisture content and temperature of materials, but the present model was simplified by applying constant parameters to confirm whether the transient temperature behavior described in the above section can be explained in terms of the moisture content dependence of microwave absorption. The scheme of this model is drawn in Fig. 9. The conservation equations are expressed by:

$$\frac{\partial T^*}{\partial \theta} = \frac{\partial^2 T^*}{\partial x^{*2}} + Q^* \tag{1}$$

$$\frac{\partial C^*}{\partial \theta} = \frac{1}{Le} \frac{\partial^2 C^*}{\partial x^{*2}} \tag{2}$$

where Le denotes Lewis number and θ is a dimensionless time defined by the Fourier number. Once the moisture content C^* achieves the equilibrium moisture content, the heat transfer is taken into account separately in two layers of wet and dry zones while the mass transfer equation is given only in the wet zone. In this analysis the equilibrium moisture constant is set to zero since it is negligibly small. Then the heat transfer equation in the dry zone is:

$$\frac{1}{\phi_1} \frac{\partial T^*}{\partial \theta} = \frac{\partial^2 T^*}{\partial x^{*2}} \tag{3}$$

where ϕ_1 is a correction factor defined by the ratio of thermal diffusivity in the dry zone layer to the wet zone layer. The dimensionless temperature and moisture content are defined by normalization by the respective initial value. The initial conditions are assumed to be uniform profiles of temperature and moisture content as:

$$T^* = C^* = 1 \tag{4}$$

Introducing the Lewis correlation, that is, the ratio of the heat transfer coefficient to the mass transfer coefficient for vapor transfer on the wet surface, which can be approximated by the specific heat of water, the boundary conditions at the wet surface of the layer are:

$$\frac{dT^*}{dx^*} = Bi\left(T^* - T_a^*\right) + \frac{r_w}{c_{pw}T_0} Bi\left(H_s - H_a\right) \tag{5}$$

$$\frac{dC^*}{dx^*} = \frac{c_p \rho}{c_{pw} C_0} Bi \cdot Le \, (H_s - H_a) \tag{6}$$

where Bi is the Biot number. After the dry surface of the layer appears, it is modeled that the interface between dry and wet zones exists inside the layer and the dry zone is developing toward to the bottom with evaporation at the interface. Then the boundary condition for the heat transfer on the dry surface exposed to air is given by

$$\frac{dT^*}{dx^*} = \frac{Bi}{\phi_2} (T^* - T_a^*) \tag{7}$$

In this case, the boundary conditions at the interface between the dry and wet zones should be necessary additionally as

$$\left. \frac{dT^*}{dx^*} \right|_{wet} = \phi_2 \left. \frac{dT^*}{dx^*} \right|_{dry} + \frac{r_w}{c_{pw} T_0} Bi \cdot f_v (x_e^*) (H_e - H_a) \tag{8}$$

$$\left. \frac{dC^*}{dx^*} \right|_{wet} = \frac{c_p \rho}{c_{pw} C_0} Bi \cdot Le \cdot f_v (x_e^*) (H_e - H_a) \tag{9}$$

where ϕ_2 is the ratio of heat conductivity in the dry zone to the wet zone, and f_v is the ratio of the overall mass transfer coefficient at the interface to that on the surface, which is given by a function of the depth of the interface from the surface x_e^*, approximated here by

$$f_v(x_e^*) = 1 - \varphi x_e^* \tag{10}$$

The boundary conditions at the bottom of the layer are

$$\frac{dT^*}{dx^*} = \frac{dC^*}{dx^*} = 0 \tag{11}$$

The heat generation rate by microwave in the wet zone is modeled approximately by a linear function of the local moisture content as (Di et al. 2000)

$$Q^* = Q_0^* C^* \tag{12}$$

while that in the dry zone is ignored.

Figure 10 shows an example of numerical results on transient profiles of moisture contents and temperatures in the layer. The thermal properties or heat transfer diffusivities are assumed to be the same in both dry and wet zones and the correction factors ϕ_1 and ϕ_2 are given as unity to make the problem simple. This assumption will not be essential for the objective of this numerical analysis. The surface of the layer is positioned at $x^* = 0$ and the bottom is at $x^* = 1$. This analysis is performed with a large Lewis number or a small moisture diffusivity to create a steep gradient of moisture content profile and to influence the internal heating rate. Thus, the initial moisture remains almost at the bottom while the surface moisture content falls rapidly and the surface dries up at $\theta = 2$. The temperature profiles are rather more uniform than the moisture content until $\theta = 0.5$ but the temperatures inside the layer to become gradually greater toward the bottom due to the internal heating effect. The internal temperatures begin falling inversely due to reduction of heating rate once the interface of wet and dry zones moves to the inside while the surface temperatures hold in a small range of 1.18–1.19 at the drying time $\theta = 2$–5. If the surface temperatures are observed carefully, there is only a momentary period of reducing temperature. However, the obvious trend of the temperature behavior during microwave drying

Fig. 10 Transient profiles of moisture content and temperature analyzed. (**a**) Moisture content profile and (**b**) Temperature profile

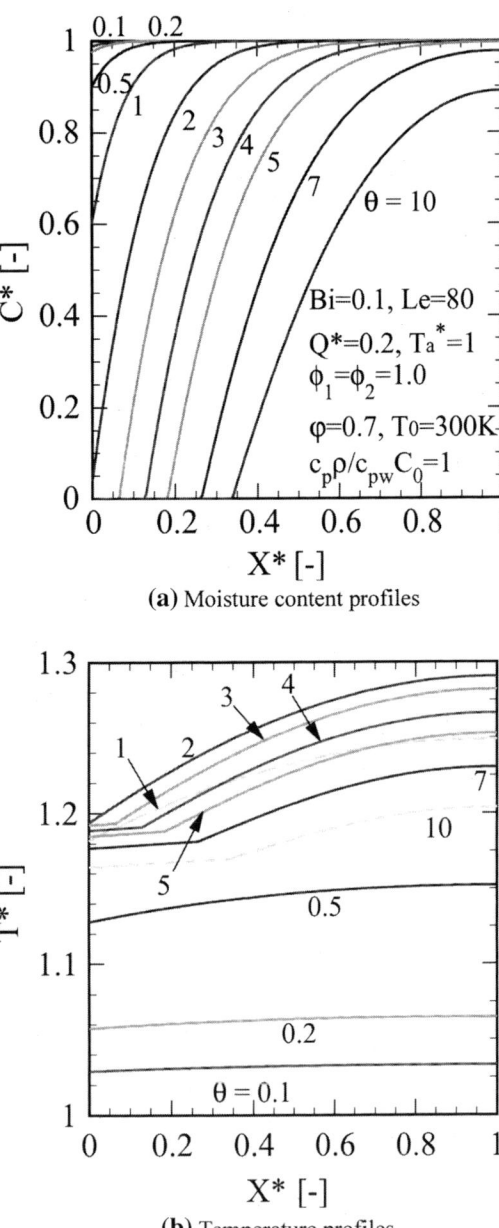

(**a**) Moisture content profiles

(**b**) Temperature profiles

seen in the previous section is not observed in the present numerical analysis. This means that the transient temperature in Fig. 6 cannot always be predicted adequately by such a simple model as assigning a linear heat generation rate to the moisture content and constant properties with respect to internal heat and mass transfer. The significant drop in temperatures in Fig. 6 may have been caused by the reason that the microwave absorption does not only decrease due to dry up around the surface but also the drying rate is accelerated due to a rise in the internal temperature and/or the

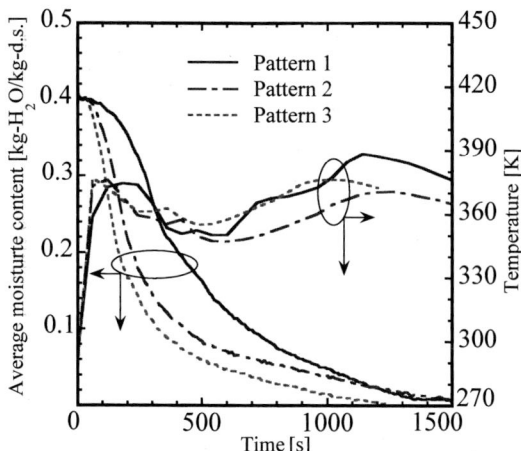

Fig. 11 Transient behavior of kaolin slab under dynamically controlled microwave heating. Pattern 1: 200 W (530 s) – 100 W, Pattern 2: 400 W (290 s) – 200 W (430 s) – 100 W, Pattern 3: 600 W (210 s) – 200 W (390 s) – 100 W

vapor pressure. Additionally, it is considered that the impedance or the interference of microwave at the test section could be shifted, as described earlier.

7 Dynamic control of microwave drying

It is implied that a rise in the internal temperature up to the boiling temperature of water causes cracking due to expansion. In order to complete microwave drying of molded samples without crack generation in as short time as possible, control of the internal temperature must be essential. In the present paper, microwave power is controlled dynamically during the progress of drying. The microwave drying experiments are performed in three dynamic patterns of power: (1) 200 W for the first 530 seconds and subsequently maintaining 100 W until the drying completion, (2) 400 W for the first 290 seconds, 200 W for the next 430 seconds and 100 W, and (3) 600 W for the first 210 seconds, 200 W for the next 390 seconds and 100 W. The time behaviors of the moisture content and the internal core temperature are plotted for each dynamic heating pattern in Fig. 11. The internal temperatures are maintained under the boiling point except in the last period of drying for any control pattern. No crack generation occurs on the sample through the drying period and the drying completes. The drying time is reduced remarkably to approximately 1500 seconds for the first and second patterns while it was the order of 10^4 seconds in the oven dryer and over 4000 seconds in hot air drying. Particularly, the third pattern is only 1250 seconds. From these facts, it is found that dynamic microwave heating contributes greatly to drying enhancement of molded ceramics.

8 Conclusions

Drying of internally heated ceramic slabs was studied to evaluate the enhancement effect and the behavior using theoretical modeling and microwave drying experiments. The experiments were carried out for a sample of kaolin molded into a slab. The results can be summarized as follows:

The theoretical analysis revealed that internal heating was the more effective method for drying enhancement while restraining a rise of internal stress generation than continuous or intermittent convective drying.

The transient behavior of microwave drying was significantly different from conventional drying by hot air heating or radiative heating in an oven dryer, and a period of momentary temperature drop in the progress of drying was observed. This phenomenon could not be predicted adequately by a simple one-dimensional model. It may occur due to the combined interactions with drying enhancement by rise in internal vapor pressure and shift of impedance or interference in the applicator. Cracking was generated by expansion due to the internal vapor pressure when the internal temperature rose up to the boiling point of water while drying by convective or radiative heating caused cracking on the sample surface due to drying induced strain–stress in the earlier drying period. Microwave drying was greatly enhanced by an internal heating effect if microwave power was controlled dynamically so as to keep the internal temperature less than 370 K. The microwave drying by the dynamic control completed successfully without crack formation in a remarkably shorter time than drying by convective or radiative heating.

References

Araszkiewicz, M., Koziol, A., Oskwarek, A., Lupinski, M.: Microwave drying of porous materials. Dry. Technol. **22**(10), 2331–2341 (2004)

Di, P., Chang, D.P.Y., Dwyer, H.A.: Heat and mass transfer during microwave steam treatment of contaminated soils. J. Environ. Eng. **126**(12), 1108–1115 (2000)

Gong, Z.X., Mujumdar, A.S., Itaya, Y., Mori, S., Hasatani, M.: Drying of clay and nonclay media: Heat and mass transfer and quality aspects. Dry. Technol. **16**(6), 1119–1152 (1998)

Islam, Md. R., Ho, J.C., Mujumdar, A.S.: Simulation of liquid diffusion-controlled drying of shrinking thin slabs subjected to multiple heat sources. Dry. Technol. **21**(3), 413–438 (2003)

Itaya, Y., Taniguchi, S., Hasatani, M.: A numerical study of transient deformation and stress behavior of a clay slab during drying. Dry. Technol. **15**(1), 1–21 (1997)

Itaya, Y., Mori, S., Hasatani, M.: Effect of intermittent heating on drying-Induced strain–stress of molded clay. Dry. Technol. **17**(7&8), 1261–1271 (1999)

Itaya,Y., Okouchi, K., Mori, S.: Effect of heating modes on internal strain–stress formation during drying of molded ceramics. Dry. Technol. **19**(7), 1491–1504 (2001)

Itaya, Y., Uchiyama, S., Cabrido, E.F., Hatano, S., Mori, S.: Uniformity of microwave field intensity by random reflection in a fluidized bed of electrically conductive beads. Kagaku Kogaku Ronbunshu (J. Chem. Eng. in Japanese) **29**(3), 339–344 (2003)

Kowalski, S.J., Rybicki, A.: Qualitative aspect of convective and microwave drying of saturated porous materials. Dry. Technol. **22**(5), 1173–1189 (2004)

Kowalski, S.J., Rajewska, K., Rybicki, A.: Mechanical effects in saturated capillary-porous materials during convective and microwave drying. Dry. Technol. **22**(10), 2291–2308 (2004)

Lehne, M., Barton, G.W., Langrish, T.A.G.: Comparison of experimental and modelling studies for the microwave frying of ironbark timber. Dry. Technol. **17**(10), 2219–2235 (1999)

Liu, F., Turner, I.W., Bialkowski, M.E.: A finite-difference time-domain simulation of the power density distribution in a dielectric loaded microwave cavity. J. Microwave Power Electromag. Energy **29**(3), 138–148 (1994)

Perre, P., Turner, I.W.: A complete coupled model of the combined microwave and convective drying of softwood in an oversized waveguide. Proceedings of the 10th International Drying Symposium (IDS'96), Krakow, Poland, vol. A, pp. 183–194 (1996)

Turner, I.W., Jolly, P.G.: The modelling of combined microwave and convective drying of a wet porous material. Drying Technology **9**(5), 1209–1270 (1991)

Turner, I.W.: A study of the power density distribution generated during the combined microwave and convective drying of softwood. Proceedings of the 9th International Drying Symposium (IDS'94), Gold Coast, Australia, pp. 89–111 (1994)

Transp Porous Med (2007) 66:43–58
DOI 10.1007/s11242-006-9021-3

ORIGINAL PAPER

The vapour–liquid interface and stresses in dried bodies

Stefan Jan Kowalski · Andrzej Rybicki

Received: 20 November 2005 / Accepted: 26 March 2006 /
Published online: 28 September 2006
© Springer Science+Business Media B.V. 2006

Abstract The paper presents a contribution to modelling the problem of vapour–liquid interface receding into dried body and stresses induced by drying of capillary-porous bodies. A complex algorithm comprising the specific mechanisms of drying in the first and second periods of drying is constructed. It enables calculation and drawing of the body temperature and drying curves for the whole drying process and identification of the vapour–liquid interface receding into the body. The drying induced stresses caused by the receding vapour–liquid interface and the non-uniform distribution of moisture content and/or temperature are analyzed. Numerical calculations of the temperature and drying curves and the drying induced stresses are carried out for the example of a finite dimensional kaolin cylinder dried convectively.

Keywords Convective drying · Vapour–liquid interface · Drying induced stresses · Modelling · Numerical calculations

Nomenclature
a Reciprocal of relaxation time [1/s]
A Elastic bulk modulus [MPa]
c_v Specific heat [J/kg · K]
c_T Coefficient of thermodiffusion [m²/s² · K]
c_X Coefficient of diffusion [m²/s²]
D_T Thermal diffusivity [m²/s]
e_{ij} Strain deviator [1]
$f(X^l)$ Moisture transport parameter [kg · s²/m⁴]

S. J. Kowalski (✉) · A. Rybicki
Institute of Technology and Chemical Engineering, Poznań University of Technology, pl. Marii Skłodowskiej-Curie 2, 60-965 Poznań, Poland
e-mail: Stefan.J.Kowalski@put.poznan.pl

A. Rybicki
e-mail: Andrzej.Rybicki@put.poznan.pl

g_i	Gravity acceleration [m/s^2]
H	Cylinder height [m]
l	Latent heat of evaporation [J/kg]
K	Elastic volumetric modulus [MPa]
M	Elastic shear modulus [MPa]
p	Pressure [Pa]
q	Heat flux [W/m^2]
r, R	Cylinder radius [m]
$\Re$	Specific gas constant [J/kg · K]
s, s^α	Entropy, entropy of α-constituent [J/kg · K]
s_{ij}	Stress deviator [Pa]
t	Time [s]
T	Temperature [K]
$\boldsymbol{u}(u_r, u_z)$	Displacement vector [m]
$\boldsymbol{W}^\alpha$	Mass flux of α-constituent [kg/m^2· s]
x_a, x_n	Mole fractions of vapour in air [1]
x, y, z	Spatial Cartesian co-ordinates [m]
X^α	Dry basis content of α-constituent [1]

Greek symbols

α_m	Coefficient of convective vapour exchange [kg · s/m^4]
α_T	Coefficient of convective heat exchange [W/m^2· K]
κ	Ratio of vapour chemical potential to that of liquid [1]
κ_v	Viscous bulk modulus [Pa · s]
$\kappa^{(T)}$	Coefficient of thermal expansion [1/K]
$\kappa^{(X)}$	Coefficient of humid expansion [1]
ε_{ij}	Strain tensor [1]
ε	Volumetric strain [1]
σ_{ij}	Stress tensor [Pa]
σ	Spherical stress [Pa], surface tension [N/m]
ρ, ρ^α	Mass density, mass concentration of α-constituent [kg/m^3]
$\hat{\rho}^\alpha$	Rate of α-constituent mass change by phase transitions [kg/m^3· s]
μ^α	Chemical potential of α-constituent [J/kg]
η	Viscosity [Pa · s]
ω	Phase transition coefficient [kg · s/m^5]
$\vartheta = T - T_0$	Relative temperature [°C]
$\theta = X^l - X_0^l$	Relative moisture content [1]
Λ^α	Mass transport coefficient of α-constituent [kg · s/m^3]
Λ_T	Coefficient of thermal conductivity [W/m · K]

1 Introduction

It is known that the classical convective drying process consists of the constant (first) and the falling (second) drying rate periods (e.g. Kudra and Strumiłło, 1986). These periods differentiate from each other in mechanisms of drying even if the external drying conditions are stable throughout the whole process. In the first period (I), the main evaporation of the moisture takes place at the body surface where it is transported from the body interior due to capillary (or other forces). The evaporation rate in this period is independent of time and is similar to that from an open liquid surface. The body temperature is constant and mostly equal to the wet bulb temperature. The first period ends at the moment when the moisture content at the body surface reaches the critical value.

Further drying drives the liquid meniscus into the body, first in larger capillaries and later in smaller ones. The rate of evaporation decreases at this stage of drying and the temperature of the body rises above the wet bulb temperature, approaching the temperature of the ambient air. During this period, termed the falling rate or the second period (II) of drying, the vapour–liquid interface recedes towards the body interior, in particular when the body is non-hygroscopic.

The problem of the vapour–liquid interface receding due to drying was discussed by several authors (e.g. Hartley 1987; Scherer 1986), but mathematical identification of this interface moving in time is poorly presented (see e.g. Luikov 1975). One can find a number of papers concerning moving boundary problems as they occur due to melting, solidification or freezing of matter, but not caused by drying. This paper is a contribution to this problem in the drying area.

Separate modelling of the individual drying periods is much easier than modelling the drying process as a whole. However, construction of a coherent model comprising both the first and second period of drying is necessary if, we want to optimize drying processes using computer simulations. Such a simulation helps us to construct processes optimized with respect to drying time. It also enables the analysis of energy consumption as well as the strength of dried materials by stress analysis (Mujumdar 2004).

The drying model comprising both drying periods becomes highly non-linear and the physical coefficients depend strongly on the moisture content and the temperature, that is, the parameters that vary in the course of drying. In this paper, such a coherent model was used to describe the theoretical drying curves and the evolution of the body temperature throughout the whole drying process and the calculation of the drying induced stresses. It also enables identification of the receding vapour–liquid interface. We illustrate the application of this model, taking as our example a cylindrically shaped kaolin sample dried convectively.

2 Mathematical modelling of drying process

The construction of the complete drying model is based on the thermomechanical theory of drying presented in Kowalski (2003). The coefficients appearing in the model equations are in general functions of moisture content and temperature. The model variant applied here is constructed under the following assumptions:

(1) The dried material is an isotropic capillary-porous body containing inside it an immiscible liquid–gas (vapour) mixture.
(2) The body is deformable, but the influence of the body's deformation on the heat and mass transfer is neglected.
(3) The phase transition of liquid into vapour proceeds close the equilibrium state both on the body surface during the first period as well as inside the body during the second period of drying.

(4) The heat necessary for evaporation of moisture is supplied from the ambient medium through the body surface. A volumetric heat supply as, for example, microwave drying, is not considered.

(5) The material coefficients existing in the equations are, in general, functions of the thermodynamic parameters, but the gradients of these coefficients are considered to be very small.

In order to develop the differential equations describing the distribution of moisture in a dried body, the following mass balance equations for the solid skeleton and the moisture are needed.

$$\dot{\rho}^s + \rho^s \mathrm{div}\dot{\boldsymbol{u}} = 0, \qquad \dot{\rho}^\alpha + \rho^\alpha \mathrm{div}\dot{\boldsymbol{u}} = -\mathrm{div}\boldsymbol{W}^\alpha + \hat{\rho}^\alpha, \qquad (1)$$

where ρ^s and ρ^α denote the mass concentration (mass per unit volume) of the solid skeleton and the moisture, with $\alpha = l$ (liquid) and v (vapour), $\boldsymbol{u}$ is the displacement vector of the solid skeleton, $\boldsymbol{W}^\alpha$ is the moisture flux with respect to skeleton (amount of moisture per unit area and unit time) and $\hat{\rho}^\alpha$ denotes the rate of mass change for constituent α due to phase transition. A dot over a symbol denotes the time derivative.

Introducing the moisture content referred to the mass of dry body defined as $X^\alpha = \rho^\alpha/\rho^s$, we can rewrite the mass balance equation for moisture as follows:

$$\rho^s \dot{X}^\alpha = -\mathrm{div}\boldsymbol{W}^\alpha + \hat{\rho}^\alpha. \qquad (2)$$

Based on the thermodynamic approach, it was concluded (see Kowalski 2003) that the moisture flux is proportional to the gradient of moisture potential μ^α, which is a function of local parameters of state such as: temperature T, strain of the porous body ε and the moisture content X^α. In accordance with the assumptions formulated above, we neglect the influence of strain on moisture transport and write

$$\boldsymbol{W}^\alpha = -\Lambda^\alpha \mathrm{grad}\mu^\alpha(T, X^\alpha), \qquad (3)$$

where $\Lambda^\alpha(T, X^\alpha)$ is termed the moisture transport coefficient.

The rate of mass change due to phase transition is found to be proportional to the difference of moisture potential of liquid and vapour, that is, (see Kowalski 2003)

$$\hat{\rho}^l = -\hat{\rho}^v = -\omega(T, X^l)(\mu^l - \mu^v), \qquad (4)$$

where $\omega(T, X^l)$ is the phase transition coefficient dependent on temperature and moisture content in the form of liquid.

It is obvious that the notion "dry body" means an absence of moisture in liquid form. Therefore, the determination of the ebbing of liquid content in the pores of a dried body is the main task for numerical calculation in drying. For this reason, we simplify the system of equations, reducing the equation for vapour content. To this end, we recall the assumption stating that the phase transitions proceed close thermodynamic equilibrium, which means that the temperatures, the pressures and the chemical potentials of liquid and vapour differ only slightly from each other. Then, we can write

$$\mu^v = \kappa \mu^l, \qquad (5)$$

where $\kappa \geq 0$ is the ratio of μ^v to μ^l (parameter of phase transition efficiency).

Based on the rate Eqs. 3 and 4 and the relation (5) and making use of the developed form of liquid potential $\mu^l(T, X^l)$ with respect to parameters T and X^l, we can write

the equation for liquid content change in the form

$$\rho^s \dot{\vartheta}^l = \Lambda^l(T, X^l)\nabla^2[c_T\vartheta + c_X\theta] - \omega(T, X^l)(1 - \kappa)[c_T\vartheta + c_X\theta], \tag{6}$$

where $\vartheta = T - T_0$ and $\theta = X^l - X_0^l$ denote the relative temperature and the relative moisture content, and recalling the last assumption, we can neglect the expression $\text{grad}\Lambda^\alpha \cdot \text{grad}\mu^\alpha \approx 0$.

The coefficient for liquid transport $\Lambda^l(T, X^l)$ is proposed to be of the following form (see Kneule 1970; Kowalski and Rybicki 2000)

$$\Lambda^l(T, X^l) = \frac{\sigma(T)}{\eta(T)} f(X^l), \tag{7}$$

where $\sigma(T) = 75[1 - 0.002(T - 273)]$ and $\eta(T) = \eta_0/[1 + 0.053(T - 273)]$ with $\eta_0 = 183 \cdot 10^{-5}$ [Pa·s], denote the surface tension and the liquid viscosity as a function of temperature, and $f(X^l)$ is a moisture transport parameter dependent on material structure and liquid content.

Coefficients c_T and c_X express the contribution of temperature and liquid content to potential μ^l. They can also be interpreted as thermodiffusional and diffusional coefficients contributing to the liquid motion inside the body (see Kowalski 2003).

The differential equation describing the temperature in dried body results from the balance of energy and the rate equation for heat transfer. The balance of energy reduced by application of Gibbs' identity takes the form Kowalski (2003)

$$\rho^s \dot{s} T = -\text{div}[\boldsymbol{q} \pm (T s^l \boldsymbol{W}^l + T s^v \boldsymbol{W}^v)], \tag{8}$$

where $s(T, X^l, X^v) = s^s + s^l X^l + s^v X^v$ denotes the total entropy of the dried body referred to the mass of dry body, s^α is the entropy of constituent α and $\boldsymbol{q}$ is the heat flux supplied through the body surface. The plus sign holds when the orientation of vectors $\boldsymbol{q}$ and $\boldsymbol{W}^\alpha$ is the same, otherwise a minus sign has to be taken. The volumetric heat supply is not considered in this paper.

The rate equation for heat transfer is Kowalski (2003)

$$\boldsymbol{q} = -\Lambda_T \text{grad} T \mp (T s^l \boldsymbol{W}^l + T s^v \boldsymbol{W}^v)], \tag{9}$$

where $\Lambda_T(T, X^l)$ is the coefficient of thermal conductivity.

Substituting the rate Eq. 9 into 8 and expanding the entropy function $s(T, X^l, X^v)$ with respect to the parameters of state, we get

$$\rho^s c_v \dot{T} + \rho^s T(s^l \dot{X}^l + s^v \dot{X}^v) = \text{div}[\Lambda_T \text{grad} T], \tag{10}$$

where $T(\partial s/\partial T) = c_v = c^s + c^l X^l + c^v X^v$ is the total specific heat of the dried body and c^α is the specific heat of constituent α.

Applying in Eq. 10 the mass balance Eq. 2 together with Eqs. 4 and 5, and next expanding the chemical potential μ^l with respect to parameters T and X^l and neglecting the heat variation inside the body due to divergence of mass fluxes as a small in comparison to the latent heat of evaporation due to phase transition, we obtain

$$\dot{\vartheta} = D_T \nabla^2 \vartheta - l\frac{\omega \cdot (1 - \kappa)}{\rho^s c_v}(c_T \vartheta + c_X \theta), \tag{11}$$

where $D_T(T, X^l) = \Lambda_T(T, X^l)/\rho^s c_v$ is the thermal diffusivity and $l = T(s^v - s^l)$ is the latent heat of evaporation. According to the last assumption it was considered that $\text{grad } \Lambda_T \cdot \text{grad } T \approx 0$ and $|\hat{\rho}^l l| \gg |Ts^l \text{div} \boldsymbol{W}^l| + |Ts^v \text{div} \boldsymbol{W}^v|$.

Equations 6 and 11 constitute the full system of equations for determination of temperature and liquid content distributions in a dried body, in which the phase transitions of liquid into vapour inside the body are taken into account. These equations have to be completed with the initial and boundary conditions for their explicit solution. Such conditions are formulated in Sect. 4 for a cylindrically shaped sample dried convectively (see Fig. 1).

3 Stresses induced by convective drying

The equations of a mechanistic model of drying presented in Kowalski (2003) have been used to determine the drying induced stresses. The dried material is assumed here to behave as elastic or viscoelastic (obeying the Maxwell model). The governing equations consist of the equation of equilibrium of internal forces, the physical relations and the equations describing the moisture evolution during drying.

Total stress acting in the fluid saturated porous medium consists in general of the effective stress in the skeleton and the stress in the pore fluid. Mostly, the stress deviator in the fluid is considered to be much smaller than that in the skeleton, so the stress in the fluid is usually taken to be the pore pressure. In convective drying the gradient of pore pressure is assumed to be insignificant, so the equation of equilibrium of internal forces in dried body is taken here in the form:

$$\sigma_{ij,j} + \rho g_i \approx 0, \qquad (12)$$

where σ_{ij} denotes the effective stress tensor, ρ is the mass density of the dried body and g_i is the gravity acceleration (neglected in further considerations).

The equilibrium equation has to be supplemented by the physical relation defining the material behavior under the action of temperature and moisture content fields. The physical relation for elastic materials is of the form:

$$s_{ij} = 2M e_{ij}, \qquad \sigma = K\left(\varepsilon - \varepsilon^{(TX)}\right), \qquad (13)$$

where $s_{ij} = \sigma_{ij} - \sigma \delta_{ij}$ is the stress deviator, $\sigma = \sigma_{ij}/3$ is the spherical stress, $e_{ij} = \varepsilon_{ij} - (\varepsilon/3) \delta_{ij}$ is the strain deviator, $\varepsilon = \varepsilon_{ii}$ is the volumetric strain, M and K are the elastic shear and bulk moduli, respectively.

The temperature and moisture content involve the volumetric thermal–humid strain (shrinkage or swelling strain), which is expressed as:

$$\varepsilon^{(TX)} = 3\left(\kappa^{(T)} \vartheta + \kappa^{(X)} \theta\right). \qquad (14)$$

It is seen that shrinkage (or swelling) strain in relation (14) is assumed to be proportional to the temperature and the content of liquid phase, $\kappa^{(T)}$ and $\kappa^{(X)}$ are the respective coefficients of linear thermal and humid expansion.

An equivalent physical relation for a viscoelastic material (Maxwell model) reads

$$\dot{s}_{ij} + \frac{M}{\eta} s_{ij} = 2M\dot{e}_{ij}, \qquad \dot{\sigma} + \frac{K}{\kappa_v}\sigma = K\left(\dot{\varepsilon} - \dot{\varepsilon}^{(TX)}\right). \qquad (15)$$

The new parameters in this relation are η and κ_v, which denote the viscous shear and bulk modules. The appearance of the viscous bulk modulus means that the theory also assumes the existence of viscous volumetric deformation, which seems to be justified in porous media.

Substituting physical relation (13) into the equilibrium condition (12) and applying this system of equations to cylindrical geometry, that is,

$$\varepsilon_{rr} = \frac{\partial u_r}{\partial r}, \quad \varepsilon_{\varphi\varphi} = \frac{u_r}{r}, \quad \varepsilon_{zz} = \frac{\partial u_z}{\partial z}, \quad \varepsilon = \frac{\partial u_r}{\partial r} + \frac{u_r}{r} + \frac{\partial u_z}{\partial z}, \quad (16)$$

we obtain the following system of two coupled equations for the determination of radial and longitudinal displacements u_r and u_z

$$M\nabla^2 u_r + \frac{\partial}{\partial r}\left[(M+A)\varepsilon - \gamma_T \vartheta - \gamma_X \theta\right] = M\frac{u_r}{r^2}, \quad (17a)$$

$$M\nabla^2 u_z + \frac{\partial}{\partial z}\left[(M+A)\varepsilon - \gamma_T \vartheta - \gamma_X \theta\right] = 0, \quad (17b)$$

where ∇^2 denotes the Laplace operator in cylindrical co-ordinates, $\gamma_T = 3K\kappa^{(T)}$, $\gamma_X = 3K\kappa^{(X)}$ and $3K = 2M + 3A$.

Since, no external surface forces act on the cylindrical sample, we assume zero-valued radial and longitudinal stresses on the external surfaces, that is,

$$\sigma_{rr}|_{r=R} = \left[2M\frac{\partial u_r}{\partial r} + A\varepsilon - \gamma_T \vartheta - \gamma_X \theta\right]|_{r=R} = 0, \quad (18a)$$

$$\sigma_{zz}|_{z=H} = \left[2M\frac{\partial u_z}{\partial z} + A\varepsilon - \gamma_T \vartheta - \gamma_X \theta\right]|_{z=H} = 0. \quad (18b)$$

The other two boundary conditions assume zero-valued radial and longitudinal displacements at the cylinder axis and the middle plane of the cylinder, that is,

$$u_r|_{r=0} \quad \text{and} \quad u_z|_{z=0}. \quad (18c)$$

We estimate the stresses in viscoelastic material using the mathematical analogy between viscoelastic and elastic stress–strain relations in the case of linear viscoelasticity and linear elasticity (see Kowalski 2003). Thus, having stress functions for elastic material, one can find the stresses for viscoelastic material using the convolution formula, namely

$$\sigma_{ij}^{(v)}(r,z,t) = \sigma_{ij}^{(e)}(r,z,t) - a\int_0^t \exp\left[-a(t-\tau)\right]\sigma_{ij}^{(e)}(r,z,\tau)d\tau, \quad (19)$$

where $a = M/\eta = K/\kappa_v$ is the reverse of the relaxation time, and $\sigma_{ij}^{(v)}$ and $\sigma_{ij}^{(e)}$ are the stresses in viscoelastic and elastic materials, respectively.

The circumferential and shear stresses for elastic cylinder reads

$$\sigma_{\varphi\varphi} = 2M\frac{u_r}{r} + A\varepsilon - \gamma_T \vartheta - \gamma_X \theta, \quad \sigma_{rz} = M\left(\frac{\partial u_r}{\partial z} + \frac{\partial u_z}{\partial r}\right). \quad (20)$$

For the present problem the state of stress in the cylinder is fully described by the following components of the stress tensor: $\sigma_{rr}, \sigma_{\varphi\varphi}, \sigma_{zz}, \sigma_{rz}$. As it is seen, the stresses are temperature and moisture content dependent. We determine these parameters in the next section.

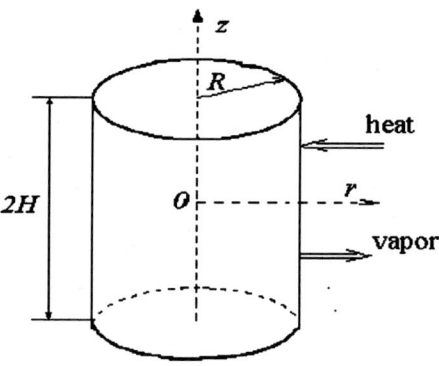

Fig. 1 Cylindrical sample under convective drying

4 Temperature and moisture content in the cylinder

The cylinder shown in Fig. 1 will be the object of our further considerations. Distribution and time evolution of liquid content and temperature will be determined for the first and second period of drying.

Differential equations 6 and 11 in cylindrical geometry take the form

$$\rho^s \dot{\theta} = \Lambda^l \left(\frac{\partial^2}{\partial r^2} + \frac{1}{r} \frac{\partial}{\partial r} + \frac{\partial^2}{\partial z^2} \right) (c_T \vartheta + c_X \theta) - \omega \cdot (1 - \kappa)(c_T \vartheta + c_X \theta), \quad (21)$$

$$\dot{\vartheta} = D_T \left(\frac{\partial^2 \vartheta}{\partial r^2} + \frac{1}{r} \frac{\partial \vartheta}{\partial r} + \frac{\partial^2 \vartheta}{\partial z^2} \right) - l \frac{\omega \cdot (1 - \kappa)}{\rho^s c_v} (c_T \vartheta + c_X \theta). \quad (22)$$

The boundary and initial conditions for mass and heat transfer are as follows:

$$\text{grad}\theta \bigg|_{\substack{r=0 \\ z=0}} = 0, \quad -\Lambda^l \text{grad}(c_T \vartheta + c_X \theta)|_{\partial B} \cdot \boldsymbol{n} = \alpha_m \left(\mu^v |_{\partial B} - \mu_a \right), \quad (23a)$$

$$\text{grad}\vartheta \bigg|_{\substack{r=0 \\ z=0}} = 0, \quad \Lambda_T \text{grad}\vartheta |_{\partial B} \cdot \boldsymbol{n} = \alpha_T (\vartheta|_{\partial B} - \vartheta_a) - l \alpha_m \left(\mu^v |_{\partial B} - \mu_a \right), (23b)$$

$$\theta(r, z, t) |_{t=0} = \theta_0 = \text{const}, \quad \vartheta(r, z, t) |_{t=0} = \vartheta_0 = \text{const}, \quad (24)$$

where $\mu^v|_{\partial B}$ and μ_a denote the chemical potentials of vapour at the boundary surface and in the ambient air (far from this surface), θ_0 and ϑ_0 are the initial moisture content and temperature, and α_m and α_T are the coefficients of the convective vapour and heat exchange between the dried body and the ambient air, respectively.

The conditions on the left-hand side of (23a) and (23b) express the symmetry in distribution of moisture content and temperature with respect to the cylinder centre, while those on the right-hand side express the convective mass and heat transfer between the cylinder boundary surface and the ambient air.

The chemical potential for vapour at the boundary surface is changed when the vapour–liquid interface recedes into the body. Solving a simple boundary value problem for vapour distribution in a single capillary tube directed in the radial direction of the cylinder, we find the following relation between the vapour chemical potential at the lateral cylinder surface and the distance r_S of the vapour–liquid interface from the cylinder centre (Kowalski 1999)

$$\mu^v|_{r=R} = \mu_n - (\mu_n - \mu_a) \ln \frac{R}{r_S} \left[\ln \frac{R}{r_S} + \frac{1}{B^v} \right]^{-1}, \qquad (25)$$

where r_S is the radius that measures the distance of the vapour–liquid interface from the cylinder centre, $B^v = \alpha_m R/\Lambda^v$ is the ratio of vapour convective exchange coefficient to the vapour transport coefficient in capillaries (Biot's number for vapour transfer) and μ_n and μ_a denote the chemical potentials of vapour in air for the saturated and unsaturated states, respectively. Note that $\mu^v|_{r=R} = \mu_n$ when $r_s = R$, $\mu^v|_{r=R}$ tends to μ_a when r_s tends to zero, and

$$\mu^v|_{r=R} - \mu_a = (\mu_n - \mu_a)[B^v \ln(R/r_S) + 1]^{-1}. \qquad (26)$$

Having the relation of type (16) for the two-dimensional case, we can reformulate the convective term for vapour transfer on the right-hand side of boundary conditions (23a) and (23b) as follows:

$$\alpha_m \left(\mu^v|_{\partial B} - \mu_a \right) = \alpha_m(\mathbf{r}_S)(\mu_n - \mu_a), \qquad (27)$$

where $\alpha_m(\mathbf{r}_S)$ is a distribution function that controls the convective vapour transfer intensity between the body and the ambient air dependent on the spatial position vector of the vapour–liquid interface $\mathbf{r}_S$ in the cylinder.

In our numerical procedure, we identify the position of vapour–liquid interface by introducing an alternative condition asking whether or not the moisture content in a given point of the cylinder reached the critical value X_{cr}. In this way we can determine the displacement of the interface with time.

In order to include or exclude the several mechanisms of drying, and in particular the term of phase transition and the moisture flux in several stages of drying, we postulate that the phase transition coefficient $\omega(T, X^l)$ (Eq. 4) and the structural moisture transport coefficient $f(X^l)$ (Eq. 7) are functions of distributive character having the following forms

$$\frac{\omega(T, X^l)}{\omega_0(T)} = \begin{cases} (X^l)^2/X_A^2 & \text{for} \quad X^l \leq X_A \\ 1 - \left(\frac{X^l - X_A}{X_{cr} - X_A} \right)^2 & \text{for} \quad X_A \leq X^l \leq X_{cr} \\ 0 & \text{for} \quad X^l \geq X_{cr} \end{cases} \qquad (28)$$

where $\omega_0(T)$ is the parameter of phase transition efficiency dependent on temperature, X_{cr} denotes the critical value of moisture content and X_A is the moisture content at the cylinder surface, being in equilibrium with the ambient air.

In the stage of drying when $X_A \leq X^l \leq X_{cr}$, the mass transport inside the body is partly in the form of liquid, and partly in the form of vapour, that is,

$$\frac{f(X^l)}{f_0} = \begin{cases} 1 & \text{for} \quad X^l \geq X_{cr}, \\ \left(\frac{X^l - X_A}{X_{cr} - X_A} \right)^2 & \text{for} \quad X_A \leq X^l \leq X_{cr}, \end{cases} \qquad (29)$$

where f_0 is a structural moisture transport parameter.

Figures 2a, b present the graphical performance of functions (28) and (29).

Function $\omega(T, X^l)$ is zero for the moisture content $X^l \geq X_{cr} \cong 0.15$ and starts to increase in the second stage of drying up to some maximum value. Below X_A the

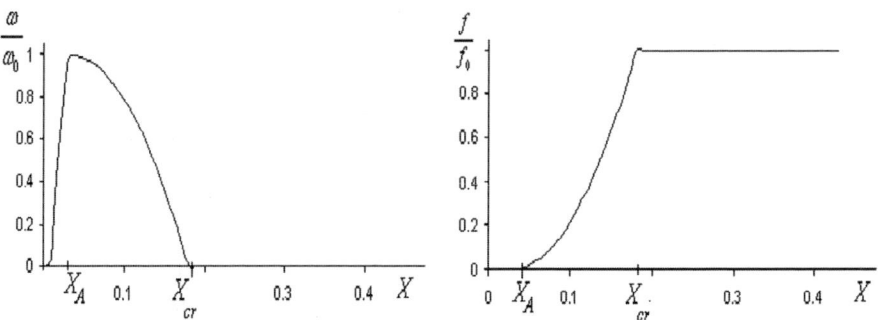

Fig. 2 Graphical performance of the distributive functions: (**a**) phase transition coefficient $\omega(T, X^l)$, (**b**) structural moisture transport coefficient $f(X^l)$

function tends to zero again. The behaviour expresses the fact that the phase transitions of liquid into vapour inside the body are neglected in the constant (I) drying rate period and are maxim during the falling (II) drying rate period when $X^l = X_A$. The intensity of phase transitions becomes zero for a dry body.

It is seen that function $f(X^l)$ approaches the maximum constant value for the moisture content $X^l \geq X_{cr} \cong 0.15$ and below X_{cr} it starts to decrease, which corresponds to the second stage of drying. Thus, it illustrates the constant and falling drying rate periods and states that liquid transport of moisture in liquid form inside the body stops totally when $X^l = X_A$.

The chemical potentials of vapour in saturated and unsaturated states of air can be written in the form (see e.g. Szarawara 1985)

$$\mu_n(p, T_n, x_n) = \mu^v(p, T_n) + \Re T_n \ln x_n, \tag{30a}$$

$$\mu_a(p, T_a, x_a) = \mu^v(p, T_a) + \Re T_a \ln x_a, \tag{30b}$$

where p is the total pressure of air, $\Re$ is the gas constant, T_n and T_a are the temperatures and x_n and x_a the mole fractions of vapour in saturated and unsaturated states of air, respectively.

Expanding potential μ^v in Taylor's series with respect to the parameters of state and using suitable thermodynamic relations for the derivatives of vapour potential, we can write the driving force for vapour transfer as follows:

$$\mu_n - \mu_a = 0.462 T_a \ln \frac{x_n}{x_a} - (7.36 - 0.462 \ln x_n)(T_n - T_a). \tag{31}$$

Such a form of driving force for vapour transfer allows us to control drying processes by a suitable choice of drying parameters T_a and x_a. The temperature T_n can be considered as wet bulb temperature and x_n as molar ratio of vapour in a saturated state of air at given drying parameters.

5 Numerical results

The numerical calculations of distributions and their evolutions in time for the temperature and the moisture content were performed for a cylindrical sample of kaolin.

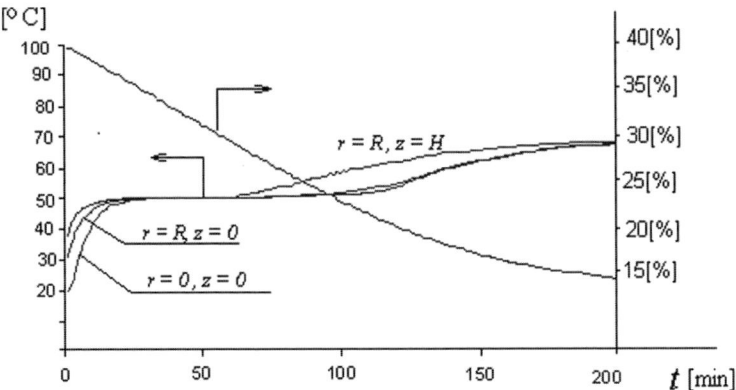

Fig. 3 Drying curve and temperature evolution curves

The following data were applied for these calculations:

$T_a = 343$ K, $T_n = 323$ K, $H = 40$ mm, $R = 40$ mm, $x_a = 0.075$,
$f_0 = 6 \cdot 10^{-7}$ kg $\cdot$ s^2/m^2, $\omega_0(1-\kappa) = 2.5 \cdot 10^{-6}$ kg $\cdot$ s/m^5, $\rho_s = 1600$ kg/m^3,
$D_T = 1.7 \cdot 10^{-3}$ m^2/s, $c_T = 3,6$ m^2/s^2 $\cdot$ K, $c_X = 10^{-3}$ m^2/s^2,
$c_v = 1.56 * 10^{-2}$ J/kg $\cdot$ K, $X_A = 5\%$, $X_0 = 40\%$, $X_{cr} = 15\%$.

Figure 3 shows the drying curve and the temperature versus time curve. The temperature of the air was assumed to be $T_a = 343$ K. The wet bulb temperature in the given drying conditions was $T_n = 323$ K. The body temperature was calculated at three points of the cylinder, namely: in the middle plane of the cylinder at r = 0, $z = 0$ and r = R, $z = 0$; and in the corner of the cylinder $r = R, z = H$.

We see that the temperatures for individual points in the middle plane of the cylinder differ only in the heating period and in the transient time between the first and second drying periods. The temperature curve for the cylinder corner starts to rise much earlier than that for the middle plane. Finally, the body temperature approaches the temperature of the ambient air. The theoretical curves in Fig. 3 reflect the experimental ones very well (see Luikov 1968, p. 151).

Figure 4 presents the distribution of temperature (temperature isolines) in the heating period (Fig. 4a), at the end of first period (Fig. 4b) and in the second period of drying (Fig. 4c). Due to symmetry, the results are presented in one quarter of the cylinder.

We see that the temperature is constant in the whole cylinder and equal to the wet bulb temperature in the constant drying rate period. In the second period the temperature is distributed throughout the cylinder. Fig. 4b shows the end of the first and beginning of the second drying period to illustrate the fact that the temperature starts to increase first in the corner of the cylinder.

Figure 5 presents the distribution (isolines) of moisture content at chosen instants of time.

The dark area represents the saturated region (overcritical moisture content, $X^l > X_{cr}$) and the pale area represents unsaturated region of the cylinder (below critical moisture content, $X^l < X_{cr}$). We see that the corner of the cylinder becomes dry first. The distribution of moisture content at 90 min. (Fig. 5b) and 120 min (Fig. 5c) of

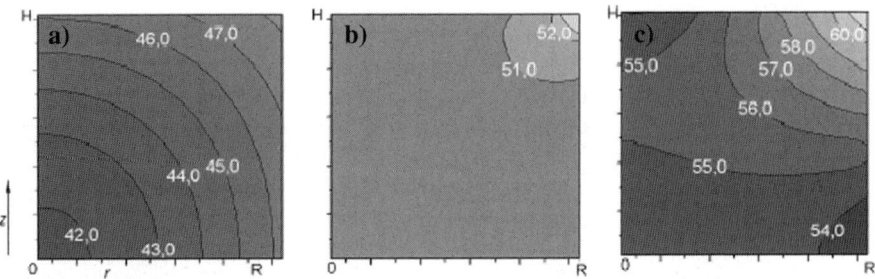

Fig. 4 Distribution of temperature [°C] in the cylinder: (**a**) heating period, (**b**) end of I period and (**c**) II period

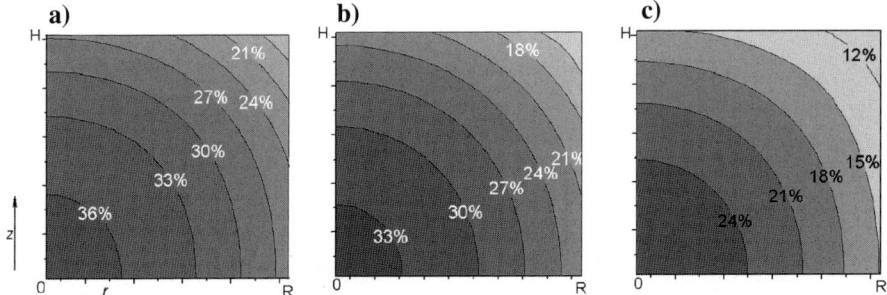

Fig. 5 Distribution of moisture content and of the vapour–liquid interface: (**a**) 60 min, (**b**) 90 min, and 120 min

drying time give an image of how the vapour–liquid interface recedes towards the interior of the cylinder in the course of drying. Note that in the present calculations the critical moisture content X_{cr} was assumed to be 15%, and this value defines the vapour–liquid interface. The moisture content decreases constantly during drying in the whole body, that is, in both saturated and unsaturated regions.

Figures 6–9 present spatial distributions of radial σ_{rr}, longitudinal σ_{zz}, circumferential $\sigma_{\varphi\varphi}$ and shear σ_{rz} stresses distributed in the quarter of the cylinder cross-sectioned with the plane (r, z).

We see that the radial stresses are equal to zero at the lateral surface of the cylinder, that is, for $r = R$ and $0 \leq z \leq H$. They are compressive in the inner part of the cylinder and become tensional close the upper part. The sign of stresses σ_{rr} changes due to the tensional stresses σ_{zz} (see Fig. 7) on the lateral surface of the cylinder $(r = R, 0 \leq z \leq H)$.

The longitudinal stresses σ_{zz} are zero on the upper surface of the cylinder i.e. for $z = H$ and $0 \leq r \leq R$. They are tensional on the lateral layer and compressive in an inner part of the cylinder.

The most dangerous stresses from the point of view of material destruction seem to be the circumferential stresses $\sigma_{\varphi\varphi}$ (Fig. 8), in particular on the lateral surface of the cylinder $r = R$, where they are tensional and reach relatively large values.

The circumferential stresses $\sigma_{\varphi\varphi}$ are tensional ones close the surface $r = R$, where the material shrinks during drying. The tensional stresses at the surface compress the interior of the cylinder so that the stresses are compressive in the cylinder core, but

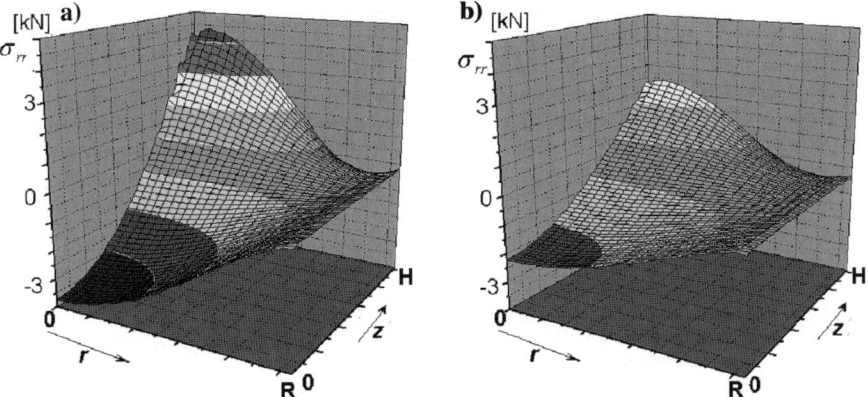

Fig. 6 Distribution of radial stresses σ_{rr} in the cylinder cross-section at 60 and 180 min drying time

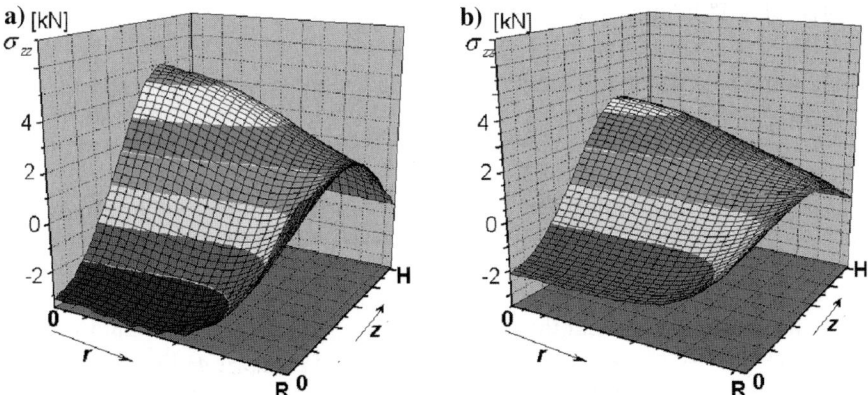

Fig. 7 Distribution of longitudinal stresses σ_{zz} in the cylinder cross-section at 60 and 180 min drying time

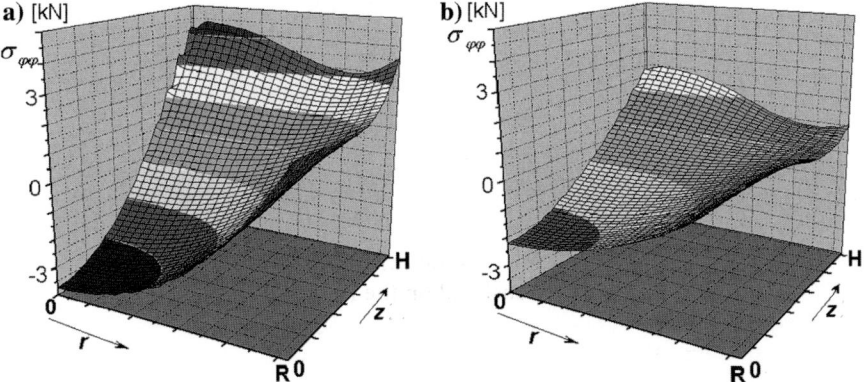

Fig. 8 Distribution of circumferential stresses $\sigma_{\varphi\varphi}$ in the cylinder cross-section at 60 and 180 min drying time

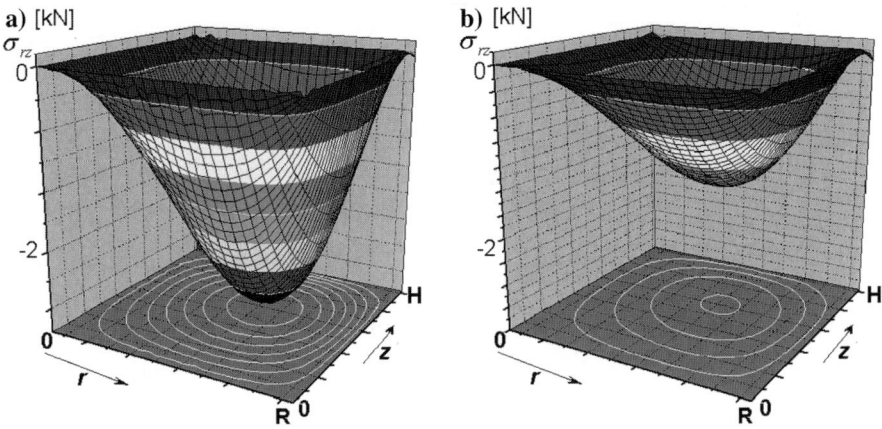

Fig. 9 Distribution of shear stresses σ_{rz} in the cylinder cross-section at 60 and 180 min drying time

only in some part of the cylinder. Similar to the radial stresses, the circumferential stresses $\sigma_{\varphi\varphi}$ inside the cylinder become tensional close to the upper part. The reasons for the sign change in this case are again the tensional stresses σ_{zz} on the lateral surface of the cylinder.

In general, the magnitude of tensional and compressive stresses has to be self-balanced in each cross-section of the cylinder, as the drying induced stresses are caused not by external forces, but by internal forces originating solely from shrinkage.

Figure 9 presents the shear stresses appearing due to non-uniform shrinkage in two spatial directions. We see that that shear stresses σ_{rz} reach their maximum values at some points inside the cylinder. Their values are not so great as the other stresses (e.g. $\sigma_{\varphi\varphi}$), but shear stresses can destroy the material structure at significantly smaller values than tensional ones. Therefore, the cracks observed inside the dried material are mostly caused by shear stresses.

Figure 10 presents the distribution of radial stresses in the middle plane of the cylinder for elastic and viscoelastic material at 120 min drying time.

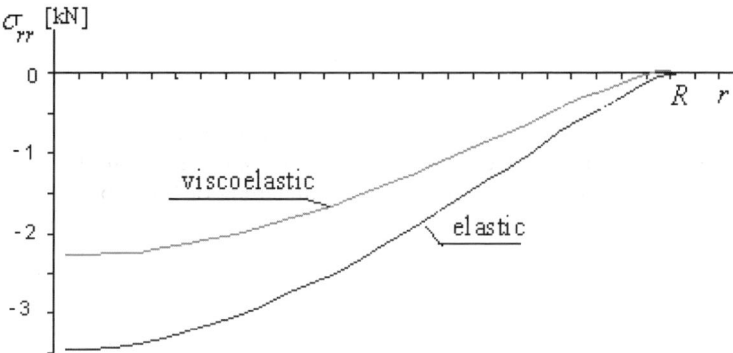

Fig. 10 Distribution of radial stresses σ_{rr} in the middle cross-section of the cylinder at 120 min drying time for elastic and viscoelastic materials

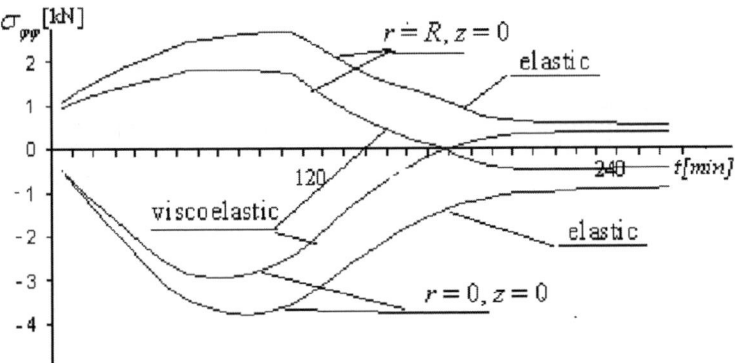

Fig. 11 Time evolution of circumferential stresses $\sigma_{\varphi\varphi}$ in the centre ($r=0, z=0$) and at the surface ($r=R, z=0$) of the cylinder for elastic and viscoelastic materials

We see an essential difference in stress distribution for elastic and viscoelastic materials as far as their value is concerned, but not only that. Figure 11 presents the time evolution of circumferential stresses in the centre ($r=0, z=0$) and at the surface ($r=R, z=0$) of the cylinder for elastic and viscoelastic material ($a = 10^{-3}$ 1/s).

It is seen that the circumferential stresses for elastic material are tensional at the surface and compressive in the core. The shape of the curves for elastic and viscoelastic cylinders is similar, but the magnitude of the circumferential stresses for the viscoelastic cylinder is significantly smaller and, furthermore, they change their sign after some time of drying. This phenomenon is termed *the stress reverse*. The stress reverse phenomenon can be explained as follows: when the cylinder dries, the drier surfaces attempt to shrink but are restrained by the wet interior. The surfaces are stressed in tension and the interior in compression and large inelastic strains occur. Later, under the surface with reduced shrinkage, the interior dries and attempts to shrink, causing the stress state to reverse (Milota and Qinglin 1994).

The smaller value of circumferential stresses for viscoelastic material follows from the relaxation phenomenon. Similar differences between stresses in elastic and viscoelastic cylinder can also be noticed for other components of the stress tensor

6 Final remarks

In the first part of this work, the coupled system of differential equations describing the heat and mass transfer during the whole drying process was developed. These equations and the numerical algorithm constructed enabled the evaluation of moisture content and temperature distribution in a dried body and their evolution in time during all stages of drying. The most relevant meaning of this model is that it permits a description of the way the vapour–liquid interface receeds into the dried body, which occurs from the moment when the moisture content at the boundary surface reaches the critical value. The results presented, in particular the drying curves and temperature plots, very satisfactorily reflect the experimental data, well known from the voluminous published literature.

The second part of this work presented the analysis of the drying induced stresses on the basis of a cylinder dried convectively. The distributions of the individual stress tensor components visualized by the surfaces created with the help of stress isolines were presented for elastic and viscoelastic cylinder based on the Maxwell model. It was stated that the stresses in viscoelastic material were smaller than in the elastic case. This follows from the relaxation phenomenon, which is typical of viscoelastic materials. Moreover, application of the viscoelastic model of the stress analysis enabled visualization of the stress reverse phenomenon, which may arise in dried material when inelastic strains occur.

The model of drying analyzed in this paper and the numerical algorithm can be used for optimization procedures and to design optimum drying processes via computer simulations.

Acknowledgements This work was carried out as a part of the research project No. 3 T09C 030 28 sponsored by Ministry of Science and Informatization in the year 2005.

References

Hartley, J.G.: Coupled heat and moisture transfer in soils: a review. Adv. Drying **4**, 199–248 (1987)
Kneule, F.: Drying, p. 363. ARKADY, Warszawa (in Polish) (1970)
Kowalski, S.J.: Thermomechanics of Drying Processes, p. 365. Springer-Verlag, Heilderberg, Berlin (2003)
Kowalski, S.J.: Theory of Flow, Heat and Diffusion Processes, p. 238. Publishers of Poznań University of Technology, Pozan (in Polish) (1999)
Kowalski, S.J., Rybicki, A.: Rate of drying and stresses in the first period of drying. Drying Technol. **18**(3), 583–600 (2000)
Kudra, T., Strumiłło, Cz.: Drying: principles, applications and design. Gordon &Brieach Science Publishers, New York (1986)
Luikov, A.W.: Systems of differential equations of heat and mass transfer in capillary porous bodies (Review). Int. J. Heat Mass Transfer **18**, 1–14 (1975)
Luikov, A.W.: Theory of Drying, Energy, p. 471 Moscow (in Russin). (1968)
Milota, M.R., Qinglin, W.: Resolution of the stress and strain components during drying of a softwood. In: Proceedings of the 9th International Drying Symposium, Gold Coast, Australia, August, 1–4, (1994)
Mujumdar, A.S.: Role of IDS in promoting innovation and global R&D effort in drying technologies. Proceedings of the 14th International Drying Symposium, pp. 22–25. S. Paulo, Brasil (2004)
Scherer, G.W.: Theory of drying. J. Am. Ceram. Soc. **73**(1), 3–14 (1986)
Szarawara, J.: Chemical thermodynamics, p. 550 WNT, Warszawa (in Polish) (1985)

Multiscale aspects of heat and mass transfer during drying

Patrick Perré

Received: 30 November 2005 / Accepted: 26 March 2006 /
Published online: 30 August 2006
© Springer Science+Business Media B.V. 2006

Abstract The macroscopic formulation of coupled heat and mass transfer has been widely used during the past two decades to model and simulate the drying of one single piece of product, including the case of internal vaporization. However, more often than expected, the macroscopic approach fails and several scales have to be considered at the same time. This paper is devoted to multiscale approaches to transfer in porous media, with particular attention to drying. The change of scale, namely homogenization, is presented first and used as a generic approach able to supply parameter values to the macroscopic formulation. The need for a real multiscale approach is then exemplified by some experimental observations. Such an approach is required as soon as thermodynamic equilibrium is not ensured at the microscopic scale. A stepwise presentation is proposed to formulate such situations.

Keywords Change of scale · Computational model · Drying · Dual scale · Homogenization · Porous media · Wood

1 Introduction

This paper focuses on multiscale modeling of coupled transfer in porous media. Nowadays, the comprehensive set of equations governing these phenomena at the macroscopic level is well known and has been widely used to simulate several configurations, particularly the drying process. However, this macroscopic description has some drawbacks: it generates a dramatic demand in physical and mechanical characterization and fails in some, not especially unusual configurations. These drawbacks are probably the main motivation for multiscale approaches. Different strategies, hence possibilities, can be applied. In the case of time scale separation, the coupling between scales

P. Perré (✉)
LERMAB (Integrated Wood Research Unit), UMR 1093 INRA/ENGREF/
University H. Poincaré Nancy I, ENGREF, 14, rue Girardet, 54 042 Nancy, France
e-mail: perre@nancy-engref.inra.fr

is sequential: the multiscale approach reduces to a change of scale. When the time scales overlap, a concurrent coupling has to be treated: this is a real multiscale configuration, more demanding in computational resources and in applied mathematics. The following content is proposed in this paper:

Sequential coupling: Techniques are available that allow macroscopic properties to be computed using the properties and morphology of the so-called unit cell. Homogenization is a part of these techniques and can be applied successfully on actual porous media such as wood, fibrous materials, solid foams, etc., provided the real morphology is taken into account. Examples of mechanical properties of oak, including shrinkage, will be considered. Finally, it has to be noted that homogenization assumes that both scales are independent, which allows the solution to be computed only once and subsequently used in the macroscopic set of equations.

Concurrent coupling: The previous assumption often fails in real life situations. In such cases, the scale level cannot be considered as independent and a multiscale approach becomes necessary. Some formulations are presented here to explain how several scales can be considered simultaneously, from a simple coupling between microscopic phases to a comprehensive formulation in which the time evolutions of the macroscopic values and macroscopic gradients are considered over the Representative Elementary Volume. Such strategies are much more demanding in terms of development and computational time. Some configurations have already been computed and are used here to picture the equations. However, the reader should be aware that this is a new and open field, especially in the domain of coupled heat and mass transfer, which is the subject of ongoing research work.

In the following, the macroscopic scale always refers to the scale we are interested in, whereas all smaller scales are referred to as microscopic scales. This indication is therefore independent of the real size of these scales. For example, when predicting shrinkage of a wood tissue, the macroscale is the cellular arrangement (typically some hundreds of micrometers) and the microscales are the cell wall (typically some micrometers) and the scale of the macromolecules (some tens of nanometers). At the opposite end, when dealing with a stack approach, the macroscale is the stack size (some meters) and the microscale is the board section (some centimeters).

2 Macroscopic formulation

Several sets of macroscopic equations are proposed in the literature for the simulation of the drying process. However, this part will just focus on the most comprehensive set of equations used at the macroscopic level, which describes the system using three independent state variables. At present, researchers using a three-variable model agree with the formulation to be used. The set of equations, as proposed below, originates for the most part from Whitaker's (1977) work with minor changes required to account for bound water diffusion and drying with internal overpressure (Perré and Degiovanni 1990). In particular, the reader must be aware that all variables are averaged over the REV (*Representative Elementary Volume*) (Slattery 1967), hence the expression "macroscopic". This assumes the existence of such a representative volume, large enough for the averaged quantities to be defined and small enough to avoid variations due to macroscopic gradients and non-equilibrium configurations at the microscopic level.

Water conservation

$$\frac{\partial}{\partial t}\left(\varepsilon_w \rho_w + \varepsilon_g \rho_v + \overline{\rho_b}\right) + \nabla \cdot \left(\rho_w \bar{\mathbf{v}}_w + \rho_v \bar{\mathbf{v}}_g + \overline{\rho_b \mathbf{v}_b}\right) = \nabla \cdot \left(\rho_g \overline{\overline{\mathbf{D}}}_{\text{eff}} \nabla \omega_v\right). \quad (1)$$

Air Conservation

$$\frac{\partial}{\partial t}\left(\varepsilon_g \rho_a\right) + \nabla \cdot \left(\rho_a \bar{\mathbf{v}}_g\right) = \nabla \cdot \left(\rho_g \overline{\overline{\mathbf{D}}}_{\text{eff}} \nabla \omega_a\right). \quad (2)$$

Energy conservation

$$\frac{\partial}{\partial t}\left(\varepsilon_w \rho_w h_w + \varepsilon_g(\rho_v h_v + \rho_a h_a) + \overline{\rho_b} \overline{h}_b + \rho_o h_s - \varepsilon_g P_g\right)$$
$$+ \nabla \cdot \left(\rho_w h_w \bar{\mathbf{v}}_w + (\rho_v h_v + \rho_a h_a)\bar{\mathbf{v}}_g + h_b \overline{\rho_b \mathbf{v}_b}\right)$$
$$= \nabla \cdot \left(\rho_g \overline{\overline{\mathbf{D}}}_{\text{eff}}(h_v \nabla \omega_v + h_a \nabla \omega_a) + \overline{\overline{\lambda}}_{\text{eff}} \nabla T\right) + \Phi, \quad (3)$$

where the gas and liquid phase velocities are given by the *Generalised Darcy Law*:

$$\bar{\mathbf{v}}_\ell = -\frac{\overline{\overline{\mathbf{K}}}_\ell \overline{\overline{\mathbf{k}}}_\ell}{\mu_\ell} \nabla \varphi_\ell, \quad \nabla \varphi_\ell = \nabla P_\ell - \rho_\ell g \nabla \chi, \text{ where } \ell = w, g. \quad (4)$$

The quantities φ are known as the phase potentials and χ is the depth scalar. All other symbols have their usual meaning.

Boundary conditions

For the external drying surfaces of the sample, the boundary conditions are assumed to be of the following form

$$\begin{aligned}\mathbf{J}_w|_{x=0^+} \cdot \hat{\mathbf{n}} &= h_m c M_v \ln\left(\frac{1 - x_{v\infty}}{1 - x_v|_{x=0}}\right), \\ P_g|_{x=0^+} &= P_{\text{atm}}, \\ \mathbf{J}_e|_{x=0^+} \cdot \hat{\mathbf{n}} &= h(T|_{x=0} - T_\infty),\end{aligned} \quad (5)$$

where $\mathbf{J}_w$ and $\mathbf{J}_e$ represent the fluxes of total moisture and total enthalpy at the boundary, respectively, x denotes the position from the boundary along the external unit normal and x_v the molar fraction of vapor.

In all these equations, subscript eff denotes the "effective" property that has to be determined experimentally or by using a predictive scaling approach (see the next section). The averaged value of variable φ, indicated by a bar, is defined as

$$\bar{\varphi} = \frac{1}{V_{\text{REV}}} \int_{\text{REV}} \varphi \, dV. \quad (6)$$

A more detailed description of these equations and related assumptions can be found elsewhere (Perré 1996, 1999). Since this formulation takes care of the internal pressure through the air balance (Eq. 2), the set of equations proved to be very powerful and able to deal with numerous configurations involving intense transfers: high-temperature convective drying, vacuum drying, RF/vacuum drying, IR/vacuum drying, etc.

For example, the simulation of convective drying at high temperature, with superheated steam or moist air, can be predicted with good accuracy when drying light concrete (Perré et al. 1993). The most important mechanisms and trends are also well predicted in the case of wood, in spite of its strong anisotropy and its biological

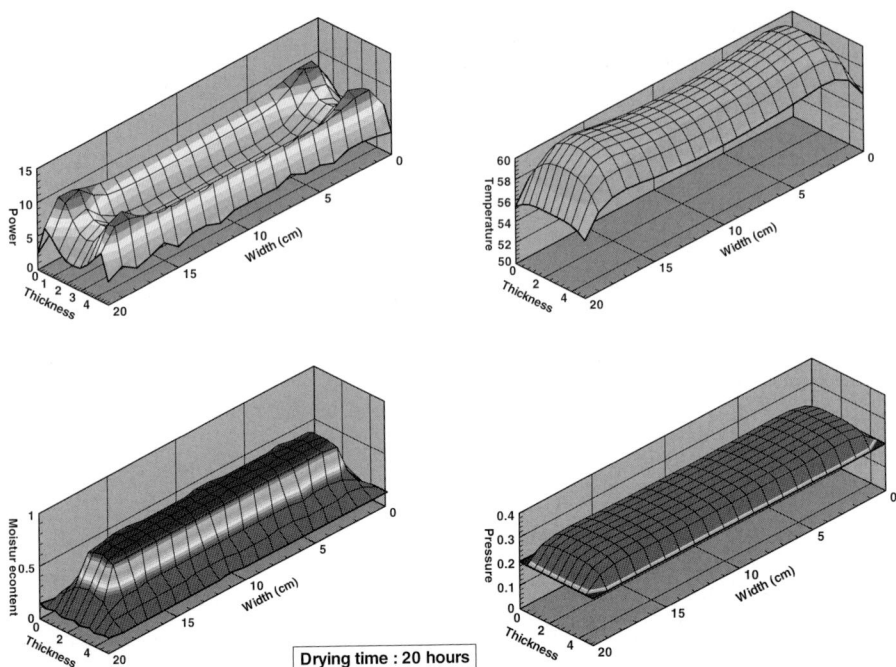

Fig. 1 Example of drying simulation: comprehensive modeling of convective drying with Radio-Frequency heating (Perré and Bucki 2004)

variability. Among the specific behaviors of wood, the internal gaseous pressure generated by internal vaporization is able to drive moisture in the longitudinal direction. This effect is easily observed and proved experimentally (by the endpiece temperature) and was simulated by using this set of equations in the drying model (Perré et al. 1993; Perré 1996).

As another example of intricate physical mechanisms, Fig. 1 depicts the variable fields (volumetric power, temperature, moisture content, and internal pressure) obtained after 20 hours of convective drying of an oak section with radio-frequency heating. In this case, the power field is computed by solving Maxwell's equations. This field depends on the dielectric properties of wood, hence on the temperature and moisture content fields. Similarly, the temperature and moisture content fields depend on the power field. To solve this two-way coupling efficiently, requires a tricky computational strategy.

3 Homogenization

Although the set of macroscopic equations presented in the previous section is a powerful foundation for computational simulation of drying, it requires knowledge of several physical parameters, most of them being a function of both temperature and moisture content. Consequently, supplying the computer model with all physical characterizations is a tedious task, which restrains the use of modeling. The first goal for a multiscale approach is to use modeling to predict one part of the parameters

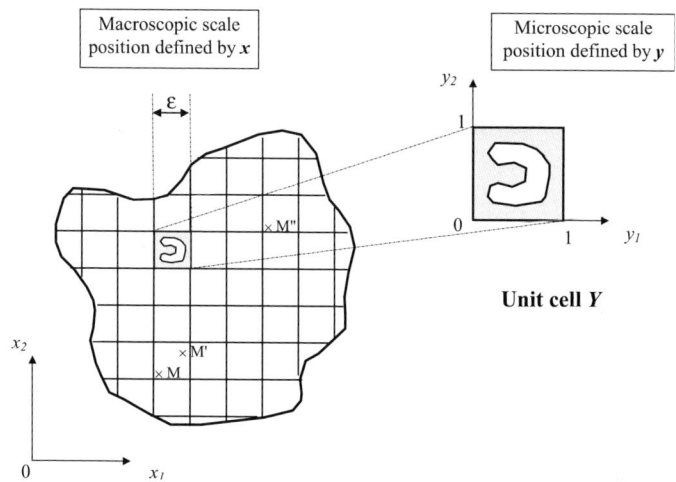

Fig. 2 Principle of double co-ordinates system used in periodic homogenization (after Sanchez-Palencia 1980)

required at the macroscale. Homogenization is one of the mathematical tools, which allows the macroscopic properties to be predicted from the microscopic description of a heterogeneous medium (Sanchez-Palencia 1980; Suquet 1985; Hornung 1997). The principle of the method will be explained hereafter, using a simple parabolic equation as reference problem:

$$\begin{aligned}\frac{\partial u}{\partial t} &= \nabla \cdot (a \cdot \nabla u) + f \quad \text{in } [0, T] \times \Omega, \\ u &= 0 \quad \text{on } [0, T] \times \partial\Omega, \\ u(0, x) &= \varphi(x) \in L^2(\Omega),\end{aligned} \quad (7)$$

where Ω is the (bounded) domain of interest, $a(x)$ the diffusion coefficient (order 2 tensor), $u(t, x)$ the variable field (i.e. temperature for thermal diffusion or moisture content for mass diffusion) f a source term and φ the initial field.

Let us now consider a heterogeneous and periodic medium, which consists of a juxtaposition of unit cells (Fig. 2). A small parameter ε denotes the ratio between the macroscopic scale, denoted by vector x, and the microscopic scale, denoted by vector y. x is used to locate the point in the macroscopic domain Ω (i.e. points M and M'' in Fig. 2) whereas y is used to locate the point within the unit cell Y (i.e. points M and M' in Fig. 2). With this new configuration, the reference problem becomes a multiscale problem:

$$\begin{aligned}\frac{\partial u^\varepsilon}{\partial t} &= \nabla \cdot \left(a^\varepsilon \cdot \nabla u^\varepsilon\right) + f \quad \text{in } [0, T] \times \Omega, \\ u^\varepsilon &= 0 \quad \text{on } [0, T] \times \partial\Omega, \\ u^\varepsilon(0, x) &= \varphi(x) \in L^2(\Omega),\end{aligned} \quad (8)$$

where $u^\varepsilon(t, x) = u(t, x, \frac{x}{\varepsilon}) = u(t, x, y)$ and $a^\varepsilon(x) = a\left(x, \frac{x}{\varepsilon}\right) = a(x, y)$, a uniformly elliptic, bounded and Y-period set of functions in $\mathfrak{R}^n$ (n = number of spatial dimensions). Although the homogenization procedure can be derived with the diffusion coefficient dependent on x, the unit cell is supposed to be dependent on y only in the following

$$a^\varepsilon(x) = a(y). \qquad (9)$$

The homogenization theory tells us that

$$u^\varepsilon \xrightarrow[\varepsilon \to 0]{} u^0 \quad \text{weakly in } H_0^1(\Omega), \qquad (10)$$

where u^0 is the solution of the homogenized problem

$$\begin{aligned} \frac{\partial u^0}{\partial t} &= \nabla \cdot \left(A^0 \cdot \nabla u^0\right) + f \quad \text{in } [0, T] \times \Omega, \\ u^0 &= 0, \quad \text{on } [0, T] \times \partial\Omega, \\ u^0(0, x) &= \varphi(x) \in L^2(\Omega). \end{aligned} \qquad (11)$$

The macroscopic property, A^0 is given by

$$A_{ij}^0 = \langle a_{ij}(y) \rangle + \sum_{k=1}^n \left\langle a_{ik}(y) \frac{\partial \xi^j}{\partial y_k}(y) \right\rangle, \qquad (12)$$

$$\underset{\text{coefficient}}{\text{Homogenized}} = \underset{\text{microscopic coefficient}}{\text{Average of the}} + \underset{\text{term.}}{\text{Corrective}}$$

In Eq. 12, the functions ξ^j are solutions of the following problems, to be solved over the unit cell Y

$$\sum_{i=1}^n \frac{\partial}{\partial y_i}\left(\sum_{k=1}^n a_{ik}(y) \frac{\partial \xi^j}{\partial y_k}(y)\right) = -\sum_{i=1}^n \frac{\partial a_{ij}(y)}{\partial y_i}, \quad j = 1,\dots,n. \qquad (13)$$

Equation 12 tells us that the macroscopic property consists of two contributions:

- the average of the microscopic properties, which accounts for the proportion and values of each phase in the unit cell;
- a corrective term, which accounts for the morphology of the constituents within the unit cell, thanks to the solution of the cell problems. This term might be very important. For example, the macroscopic stiffness of the earlywood part of softwood is only 5–10% of the averaged value of the microscopic properties, which means that the corrective term removed 90–95% of this averaged value (Farruggia 1998).

Equations 12 and 13 can be derived either by the method of formal expansion or, more rigorously, by using modified test functions in the variational form of (8) (Sanchez-Hubert and Sanchez-Palencia 1992). To derive the limit problem in a formal way, the unknown function u^ε is developed as the following expansion

$$u^\varepsilon(t, x, y) = u_0(t, x) + \varepsilon u_1(t, x, y) + \varepsilon^2 u_2(t, x, y) + \cdots \qquad (14)$$

Due to the rapid variation of properties inside Y, two independent space derivatives exist

$$\nabla(u) = \nabla_x(u) + \frac{1}{\varepsilon}\nabla_y(u). \qquad (15)$$

Applying this derivative rule to problem (8) leads to a formal expansion of the parabolic equation in powers of ε. The term with ε^{-2} tells us that u_0 does not depend on y.

Using this result, the term with ε^{-1} gives

$$\nabla_y \cdot \left(a(y)\nabla_y u_1(t,x,y)\right) = -\nabla_y \cdot (a(y)\nabla_x u_0(t,x)). \tag{16}$$

Equation 16 is linear, so u_1 can be expressed as a linear expression of the derivatives of u_0 with respect to x

$$\nabla_y u_1(t,x,y) = \sum_{j=1}^{n} \nabla_y \xi^j(y) \frac{\partial u_0(t,x)}{\partial x_j}. \tag{17}$$

In Eq. 17, $\xi^j(y) \in H^1_{\text{per}}(\Omega)$ are Y-periodic functions, solutions of the following problems

$$\nabla_y \cdot \left(a(y)\nabla_y \xi^j(t,x,y)\right) = -\nabla_y \cdot \left(a(y)\vec{e}_j\right). \tag{18}$$

Note that Eq. 18 is just Eq. 13 written with the derivative notation defined in Eq. 15. $\vec{e}_j$ is the unit vector of axis j.

Finally, the term with ε^0 reads

$$\frac{\partial u_0}{\partial t} + \nabla_x \cdot \left(a(y)\left(\nabla_x u_0 + \nabla_y u_1\right)\right) + \nabla_y \cdot \left(a(y)\left(\nabla_x u_1 + \nabla_y u_2\right)\right) = f. \tag{19}$$

Averaging Eq. 19 over the unit cell Y allows the Y-periodic terms to vanish

$$\frac{\partial u_0}{\partial t} + \nabla_x \cdot \left((\langle a(y)\rangle + \langle b(y)\rangle)\nabla_x u_0\right) = f$$
$$\text{with } b_{ij}(y) = \sum_{k=1}^{n} a_{ik}(y) \frac{\partial \xi^j(y)}{\partial y_k}. \tag{20}$$

Equation 20 is just the homogenized problem and the rule to obtain the homogenized property A^0, as already formulated in Eqs. 11 and 12, respectively.

The detail of the formal expansion in powers of ε for the mechanical problems encountered in drying can be found elsewhere: for elasticity (Sanchez-Palencia 1980; Léné 1984; Sanchez-Hubert and Sanchez-Palencia 1992), for thermo-elasticity (L'Hostis 1996) and for shrinkage (Perré 2002; Perré and Badel 2003).

The homogenization formulation results in classical PDE problems. Moreover, owing to the assumption that the microscopic and macroscopic scales are independent, together with the simple physical formulation used in this work (elastic constitutive equation and shrinkage proportional to the change of moisture content), the problems are steady-state, linear, and uncoupled.

However, the computer model must be able to handle any geometry and deal with properties that vary strongly in space. This is why the FE method is among the appropriate numerical strategies. Finally, the mesh must represent the real morphology of the porous medium as closely as possible. The best strategy able to fulfil this requirement consists in building the mesh directly from a microscopic image of the porous medium. To address this demand, two numerical tools have been developed:

(1) *MeshPore*: software developed to apply image-based meshing (Perré 2005);
(2) *MorphoPore*: code based on the well-known FE strategy, specifically devoted to solving homogenization problems, namely to deal with all kinds of boundary conditions encountered in solving the cell problems.

Figure 3 depicts one typical set of solutions, which allows the macroscopic properties (stiffness and shrinkage) of one annual ring of oak to be calculated. In this figure, the solid lines represent the initial position of boundaries between different kinds of tissues (vessels, parenchyma cells, fiber, and ray cells), while the colored zones represent the deformation of these zones as calculated for each elementary solution (an amplification factor is applied so that the deformation field can be observed easily). These solutions emphasize the complexity of the pore structure of oak, and its implication on mechanical behavior. For example, the fiber zones are strong enough to impose this shrinkage on the rest of the structure (problem w) and the ring porous zone is a weak part unable to transmit any radial forces (in this part, the vessels enlargement is obvious, problem 11) and prone to shear strain (problem 12).

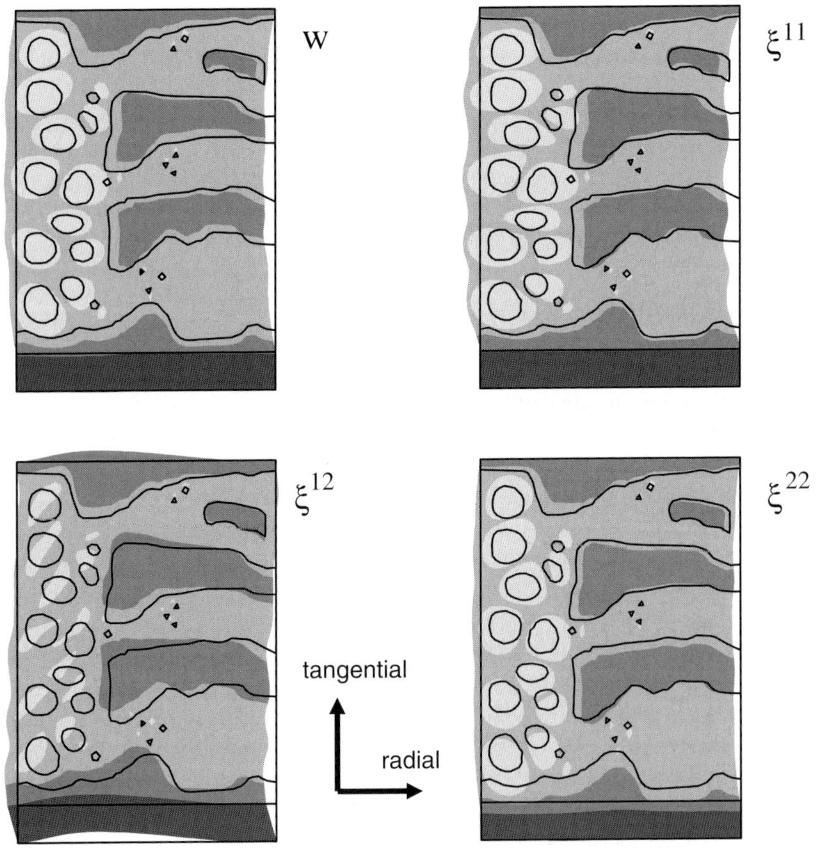

These products are written in Fortran 95, but they run on a PC as classical Windows applications, thanks to a graphical library used for pre- and post-processing (Winteracter 5.0).

Fig. 3 Periodic displacement fields computed for the four problems to be solved over the representative cell. (W) shrinkage problem ; (ξ_{11}) and (ξ_{22}) stiffness problems in radial and tangential direction. (ξ_{12}) corresponds to the shear problem. *Solid lines* represent the initial contours of tissues (Perré and Badel 2003). Color code, from *dark* to *light* : ray cells, fiber zones, parenchyma cells, and vessels

This approach allowed us to quantify the effect of fiber proportion and fiber zone shape on the macroscopic values or to predict the increase of rigidity and shrinkage coefficients due to the increase of the annual ring width (Perré and Badel 2003).

Keeping in mind that this homogenization procedure was obtained by letting ε tend towards zero, the time variable undoubtedly disappears within the unit cell Y. Consequently, the macroscopic property has to be calculated only once and subsequently used in the homogenized problem. This is a typical sequential coupling.

Up to here, we have presented a set of macroscopic equations that allows numerous transfer and drying configurations to be computed and a mathematical method, homogenization, that allows macroscopic properties (the "effective" parameters of the macroscopic set) to be computed from the constitution of the material at the microscopic level (concept of unit cell, or REV, Representative Elementary Volume).

Although consistent and comprehensive, this approach is not relevant for certain configurations. The last part of this paper is therefore devoted to concurrent coupling.

4 Dual-scale methods

Figure 4 depicts two examples for which the previous approach fails: soaking samples of hardwood with water. In such a process, a typical dual scale mechanism occurs: the water flows very rapidly in those vessels that are open and connected. In oak, this easy transfer happens in the vessels without (or with a low amount of) thyloses, whereas the early part of each annual growth ring is very active in the case of beech (Fig. 4). Then, moisture needs more time to invade the remaining part of the structure, by liquid transfer in low-permeable tissues or by bound water and water diffusion. The photographs in Fig. 4 show that the local thermodynamic equilibrium cannot always be assumed. This occurs as soon as the macroscopic and the microscopic time scales have the same orders of magnitude. As a result, the microscopic field not only depends on the macroscopic field, but also on the history of this macroscopic field. Such media manifest microscopic storage with memory effects. Due to this phenomenon, for example, transient diffusion in a sheet of paper presents a non-Fickian behavior: the characteristic time does not increase as the thickness squared (Lescanne et al. 1992). Obviously, the previous formulations (macroscopic formulation and change of scale) must be discarded.

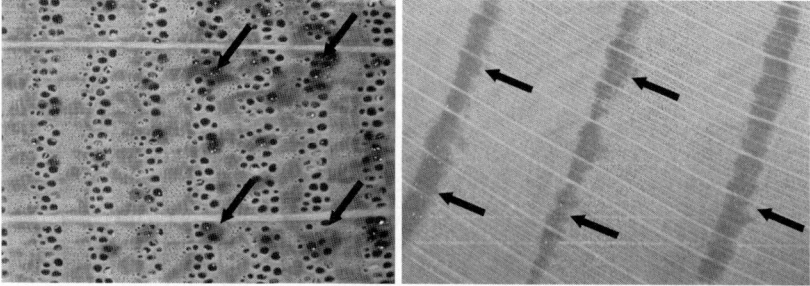

Fig. 4 Absence of local thermodynamic equilibrium when soaking wood samples, oak (*left*) and beech (*right*). View of the outlet face (S. Ghazil–LERMAB)

Different strategies have been imagined, from simple global formulations to more comprehensive ones:

(1) relatively simple global formulations are able to account for the microscopic delay but have to be fed by experimental knowledge;
(2) a mesoscopic model consists in dealing with the microscopic detail at the macroscopic level. This is just a continuous model for which the physical parameters change rapidly in space. Such models are obviously able to catch all subtle mechanisms and interaction between scales, but are very demanding in terms of computer resources. They must be limited to cases for which the ratio between the microscale and the macroscale (ε) remains close to unity;
(3) finally, homogenization can be extended to these configurations without local equilibrium. To do this, the microscopic transfer properties have to be scaled by ε^2 before calculating the limit as ε goes to zero (Hornung 1997).

This stepwise approach is presented briefly in the next sections for a simplified formulation, even though, we have to keep in mind that more comprehensive formulations are required to catch the coupling between transfer that takes place during drying.

Parallel flow models

The parallel flow model assumes that two continuous phases exist at the microscopic level. They participate in the macroscopic fluxes and can also exchange at the microscopic level.

$$\frac{\partial}{\partial t}(au_1) - \nabla \cdot (A\nabla u_1) + k(u_1 - u_2) = f_1,$$
$$\frac{\partial}{\partial t}(bu_2) - \nabla \cdot (B\nabla u_2) + k(u_2 - u_1) = f_2, \qquad (21)$$

where a and A are the accumulative and transfer coefficients for material 1, respectively, b and B the same quantities for material 2. k is an exchange coefficient between the two materials and f_1 and f_2 are the source terms for materials 1 and 2.

These equations apply for different mechanisms, for example, heat conduction, mass diffusion or liquid migration (Darcy's law), variables u_1 and u_2 being the temperature, the moisture content or the liquid pressure, respectively.

This set of equations can be considered as a macroscopic formulation in which two phases are acting in parallel and can exchange when their potentials are not identical (Fig. 5). k, the transfer coefficient, is a global parameter that needs to be identified or measured.

A simplified configuration of model (21) is often encountered in practice: one phase has low storage but high-transfer abilities, whereas the other has exactly opposite features ($a = 0$ and $B = 0$). In this case, material 1 is a conductive phase having a connected network and material 2, which ensures storage, consists of a low-conductive phase organized as isolated cells. This applies, for example, to liquid migration in fissured blocks, to heat conduction in a heat exchanger with storage capacity or moisture diffusion in a porous medium having a connected conductive phase (granular media, fibrous materials such as paper or medium density fiberboard, etc.).

Parallel flow models

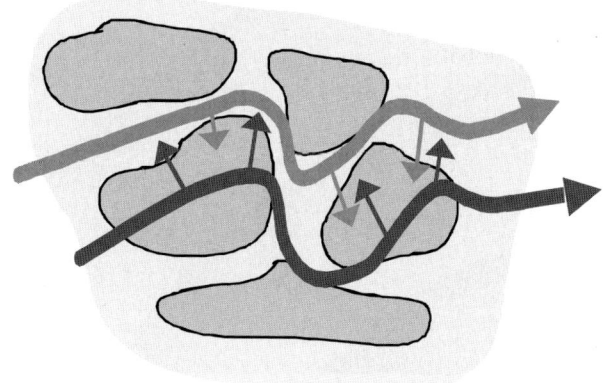

Fig. 5 Concept of parallel flow model

The mesoscopic model

The mesoscopic model simply consists of accounting for the rapid variations of the material properties within the unit cell. This strategy requires the complex geometrical structure of the heterogeneous medium to be meshed and the grid to be refined enough for the small structure to be caught. Such a strategy requires a massive amount of computational resources and works only when the size of the unit cell is just slightly smaller than the size of the domain Ω. Figures 6 and 7 depict an example of the mesoscopic approach used to simulate the drying of a section of softwood containing some annual growth rings (Perré and Turner 2002).

A similar strategy was also adopted to obtain a realistic description of fluid flow in beech. In this case the x–y plane is the unit cell (microscopic cylindrical coordinates, with the vessel in the center surrounded by the fiber zone) and the z-axis the macroscopic direction for fluid flow (Perré 2004). Figure 8 depicts, as carpet plots, the moisture content fields computed at different soaking times. A fast wetting front invades the vessel first. This phenomenon lasts about one minute. After this first step, one can see how moisture migrates in the fiber zone by two different mechanisms:

(1) A radial migration from the invaded vessel along the entire sample length (20 cm in this case);
(2) A significant macroscopic migration within the fiber zone from the injection plane (especially at 2 h). This effect could have been computed only by using a complex microstructure model (Schowalter *in* Hornung 1997).

Distributed microstructure models

Distributed microstructure models are among the multiscale models with concurrent coupling (Schowalter *in* Hornung 1997). In this section, a bounded macroscopic domain Ω is considered. This domain is supposed to be periodic and defined by a unit cell Y. This unit cell consists of a conductive phase, denoted C, and a storage phase, denoted S (Fig. 9). Here, the conductive phase is supposed to be connected at the

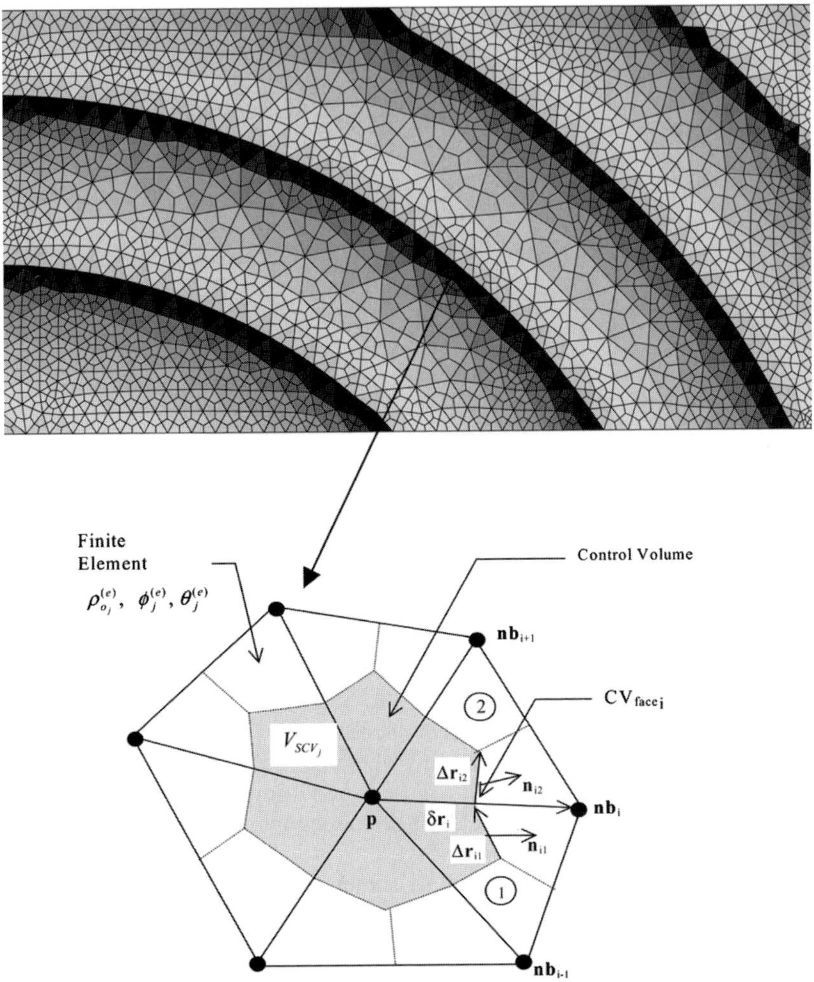

Fig. 6 Mesh structure including the FE mesh and the Control volumes built around each node. Dark shaded regions on the mesh represent latewoood (high density) and light shaded regions represent earlywood (low density) (Perré and Turner 2002)

macroscale. By scaling the transfer property by ε^2 in the storage phase, homogenization gives rise to a model in which non-equilibrium can exist within the unit cell. This results in a macroscopic equation with a delay due to the storage capacity of the unit cell

$$|C| \frac{\partial u}{\partial t} - \nabla.(A_C \nabla u(t,x)) + Q(x,t) = 0 \quad x \in \Omega,$$
$$b \frac{\partial U}{\partial t} - \nabla_y.(A_S \nabla U(t,x,y)) = 0 \quad y \in S_x, \quad x \in \Omega, \quad (22)$$

where b is the accumulative coefficient for the storage phase and $|C|$ is the accumulative coefficient of the conductive phase weighted over its volume in the unit cell Y. u is the macroscopic variable and U the microscopic variable in the storage phase S.

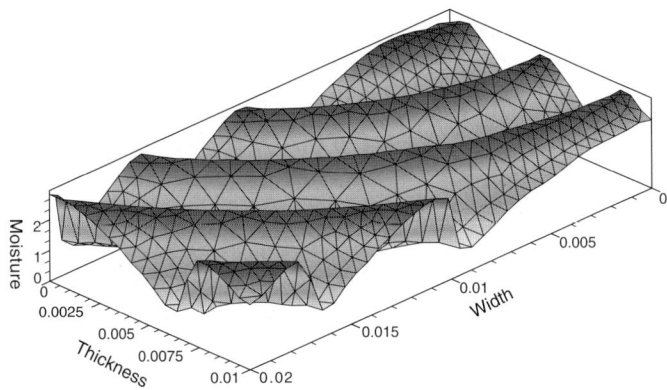

Fig. 7 Drying a softwood section using a mesoscopic model. The earlywood/latewood alternation is obvious in this initial moisture content field (Perré and Turner 2002)

These variables are connected by boundary conditions at the interface Γ between phases S and C (potential continuity and mass conservation)

$$U(t,x,y) = u(t,x) \quad y \in \Gamma, \; x \in \Omega,$$
$$Q(t,x) = \int_\Gamma A_S \nabla_y U \cdot \vec{n}\, \mathrm{d}S, \quad x \in \Omega. \tag{23}$$

Although this set resembles expression (21), two main differences have to be emphasized:

(1) the macroscopic property A_C is computed from a homogenization procedure applied to part C of the unit cell;
(2) the memory effect is not obtained thanks to an identified parameter, but from the computation of the variable field over part S of the unit cell.

If necessary, this model can be refined to account for additional phenomena (Schowalter, Arbogast *in* Hornung 1997):

(1) a linear approximation of the variable U over the boundary Γ, to account for the macroscopic gradient of variable u;
(2) a small contribution of the secondary flux in the storage phase, by adding the divergence of the microscopic flux over the S block in the source term Q;
(3) a medium contribution of the secondary flux in the storage phase, by using a additional macroscopic variable accounting for the flux inside the macroscopically connected phase S.

Computational considerations

This kind of model has been developed successfully in the case of paper (unit cell a in Fig. 9) or for a whole stack of boards (unit cell b in Fig. 9, see Perré and Rémond 2006). A more comprehensive formulation than the one presented in Eqs. 22 and 23 was used. Indeed, a 2 or 3 state variable is used at the macroscopic and the microscopic scales. One has to keep in mind that such dual scale models (concurrent coupling) are demanding in terms of computational resources, because the variable field in the microscopic unit cell has to be updated in time for each point x of the macroscopic grid.

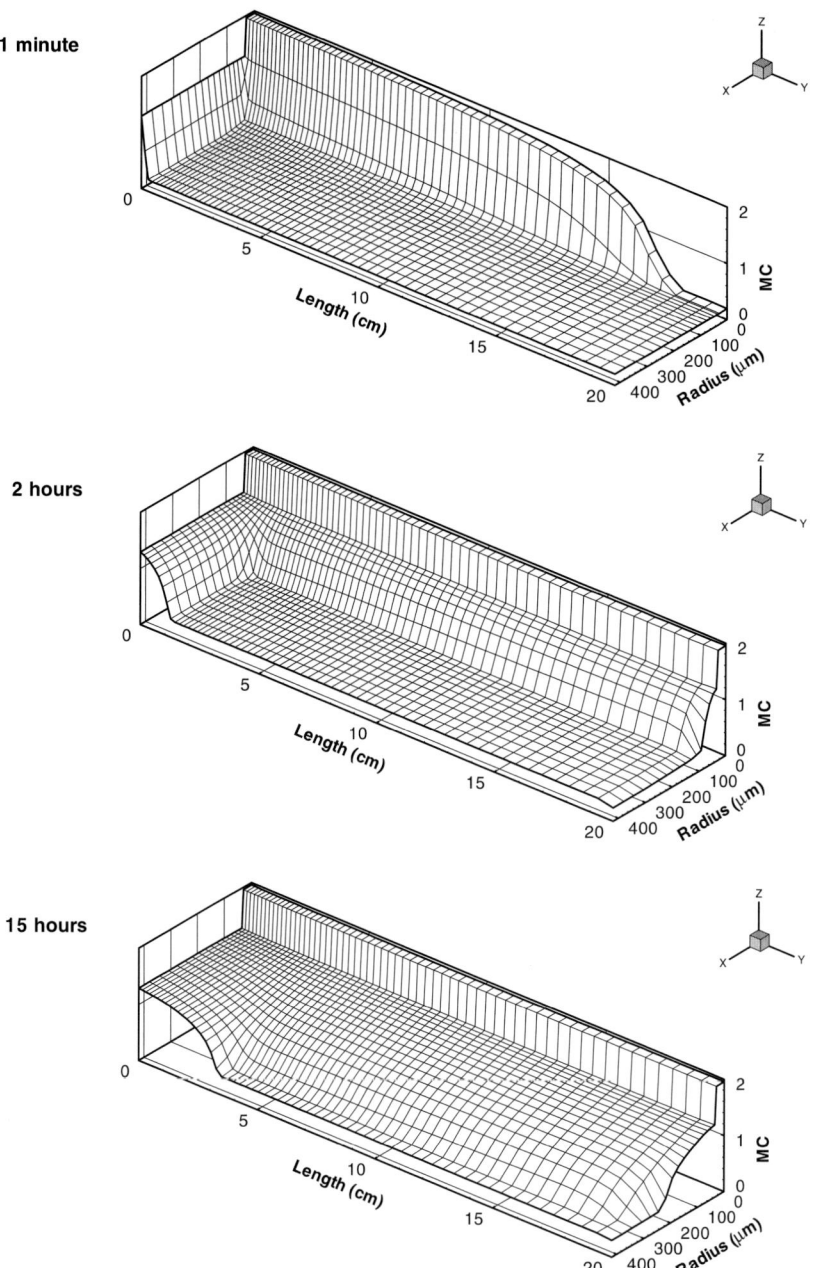

Fig. 8 Moisture content fields calculated at different re-saturation times using a mesoscopic model to simulate rapid fluid migration in the vessel and slow migration in the surrounding zone (Perré 2005)

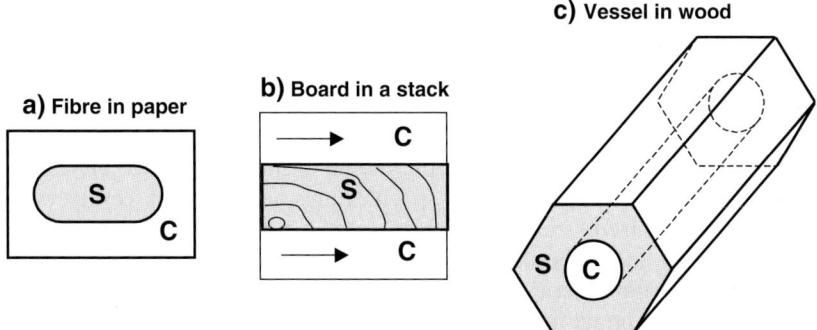

Fig. 9 Some unit cells suitable for different configurations of dual-scale modeling: S = storage phase, C = conductive phase

The coupling between the two scales is a two-way process:

(1) the source terms (heat, vapor, and liquid) crossing each surface Γ at time t, allow the set of coupled macroscopic equations to be advanced in time by a macroscopic time increment dT;
(2) new boundary conditions are use for each microscopic module to advance up to time $t + \mathrm{d}T$ by several microscopic time intervals dt.

This coupling procedure is iterated up to the end of the process. In order to be efficient, each module has its own microscopic time step, automatically adjusted during the calculation according to the convergence conditions of that unit cell. Obviously, the coupling between the two scales requires the global time step dT to be adjusted during the process. Generally, the microscopic time steps are different from one unit cell to the other and smaller than the global one. However, when the convergence conditions are favorable for one given unit cell, the microscopic time step might be larger than the global one. In this case, the microscopic drying is computed with one single time step, equal to dT. To minimize the computational time, the time step allowing convergence at the microscopic scale is stored to anticipate a possible increase of the global time step dT. For such configurations with very poor convergence conditions, a dual-scale Newton–Raphson strategy has to be adopted. This method allows the coupling between scales to be taken into account with a quadratic convergence rate.

One example: a stack/board model

Recently, the concept of the distributed microstructure model has been applied to simulate the behavior of a whole stack of boards during drying (Perré and Rémond 2006). This configuration is typical of industrial batch dryers. The boards are stacked on stickers, which preserves air canals between two adjacent board layers. Large fans force the air in each canal, which ensures heat supply and vapor exhaust to and from boards. In this dual-scale model, each micro-model simulates the coupled heat and mass transfer within one single board. The macro model deals with heat and mass transfer in each airflow canal. The coupling between scales is a two-way process that requires a refined strategy for numerical convergence:

(1) the heat and mass transfer fluxes at each board surface are involved in the airflow balance equations;
(2) the drying conditions used in each micro-model comes from the airflow properties at that board position.

This model can be used to simulate the drying of identical boards (Fig. 10). This approach allows, for example, the kiln specifications and the stack configuration to be tested and optimized (sticker thickness, air velocity, stack width etc.). But the potential of such a model is much more challenging. By its ability to simulate the drying of boards with different properties, the natural variability of wood can be accounted for in the simulation (Fig. 11). For the first time it becomes reasonable to use a computational approach to improve or imagine drying schedules. Indeed, due to wood variability, any attempt to optimize the drying schedule with a model able to simulate only one board failed. Just one example: in industry, the final moisture

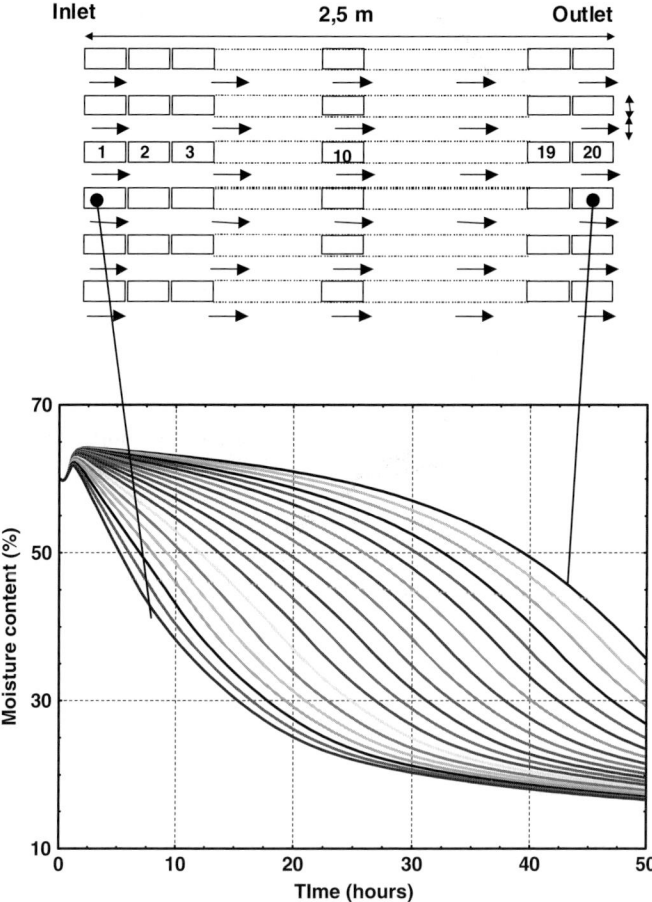

Fig. 10 The dual-scale (*stack/board*) model used to simulate the drying of identical boards: the interaction between each board and the air flow changes the drying conditions from the inlet to the outlet, which explains the difference in drying kinetics

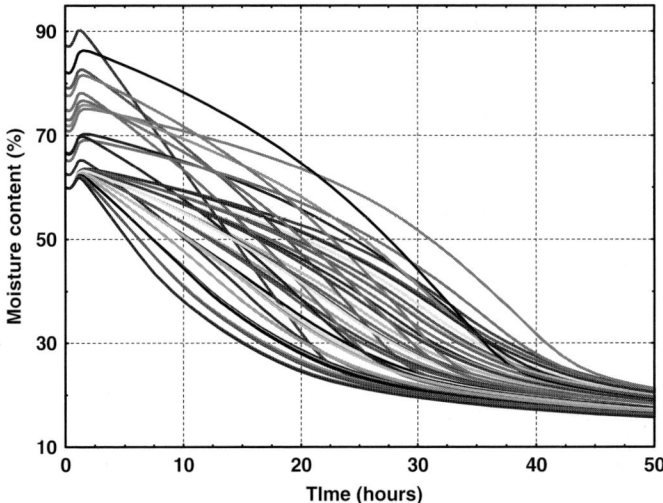

Fig. 11 The dual-scale (*stack/board*) model used to simulate the drying of boards with variable properties (transport properties and initial moisture content). The drying kinetics of each board depends on its individual properties, on its position within the stack and on the effect of all boards situated upstream

content variability within boards is of utmost importance in regard to the drying quality. Only a dual scale model can predict such a variability.

5 Conclusion

This paper gives an overview of multiscale modeling of coupled transfer in porous media. A stepwise approach is proposed, starting from the macroscopic formulation and ending with modern and powerful approaches. The difference between *change of scale* (sequential coupling between scales) and *multiscale modeling* (concurrent coupling) is clearly presented. We also emphasize the difference between formulations that need to be fed by experimental measurements or identified parameters and approaches that permit, somehow, the macroscopic properties to be predicted from the so-called unit cell (morphology and microscopic properties).

References

Farruggia, F.: Détermination du comportement élastique d'un ensemble de fibres de bois à partir de son organisation cellulaire et d'essais mécaniques sous microscope. Thèse de Doctorat-ENGREF (1998)

Hornug, U.: Homogenization and porous media. Springer–Verlag, New York (1997)

Léné, F.: Contribution à l'étude des matériaux composite et de leur endomagement. Thèse d'état, Université Pierre et Marie Curie, Paris (1984)

Lescanne, Y., Moyne, C., Perré, P.: Diffusion mechanisms in a sheet of paper DRYING'92, pp. 1017–1026. Elsevier Science Publishers B.V., Amsterdam (1992)

L'Hostis, G.: Contribution à la conception et à l'étude de structures composites thermoélastique. Thèse de Doctorat, Université Paris 6 (1996)

Perré, P., Turner, I.W.: A Heterogeneous Wood Drying Computational Model that accounts for Material Property Variation across Growth Rings. Chem. Eng. J. **86**, 117–131 (2002)

Perré, P.: The numerical modelling of physical and mechanical phenomena involved in wood drying: an excellent tool for assisting with the study of new processes, Tutorial. In: Proceedings of the 5th International IUFRO Wood Drying Conference, pp. 9–38. Qué., Canada (1996)

Perré, P.: How to get a relevant material model for wood drying simulation ? In: Invited Conference, Proceedings of the First COST Action E15 Wood Drying Workshop, 27p. Edinburgh, Scotland (1999)

Perré, P.: Wood as a multi-scale porous medium : Observation, Experiment, and Modelling. In: Proceedings of the First International conference of the European Society for wood mechanics, pp. 365–384. EPFL, Lausanne, Switzerland (2002)

Perré, P.: *MeshPore* : a software able to apply image-based meshing techniques to anisotropic and heterogeneous porous media. Dry. Technol. J. **23**, 1993–2006 (2005)

Perré, P.: Evidence of dual scale porous mechanisms during fluid migration in hardwood species : Part II : A dual scale computational model able to describe the experimental results. Chin. J. Chem. Eng. **12**, 783–791 (2004).

Perré, P., Badel, E.: Properties of oak wood predicted from X-ray inspection: representation, homogenisation and localisation. Part II: Computation of macroscopic properties and microscopic stress fields. Ann. For. Sci. **60**, 247–257 (2003)

Perré, P., Bucki, M.: High-frequency/vacuum drying of oak : modelling and experiment. In: Proceedings of the 5th Workshop of COST Action E15, 10p. Athens, Greece (2004)

Perré, P., Degiovanni, A.: Simulations par volumes finis des transferts couplés en milieu poreux anisotropes : séchage du bois à basse et à haute température. Int. J. Heat and Mass Transfer **33**, 2463–2478 (1990)

Perré, P., Moser, M., Martin M.: Advances in transport phenomena during convective drying with superheated steam or moist air. Int. J. Heat Mass Tran. **36**, 2725–2746 (1993)

Perré, P., Rémond, R.: A dual scale (board and stack) computational model of wood drying, Drying Technology, to appear.

Sanchez-Palencia, E.: Non-homogeneous media and vibration theory. Lecture Notes in Physics, 127, Springer Verlag, Berlin (1980)

Sanchez-Hubert, J., Sanchez-Palencia, E.: Introduction aux méthodes asymtotiques et à l'homogénéisation Masson, Paris.

Slattery, J.C.: Flow of viscoelastic fluids through porous media. Am. Inst. Chem. Eng. J. **13**, 1066–1071 (1967)

Suquet, P.M.: Element of homogenization for ineslastic solid mechanics. In: Sanchez-Palencia and Zaoui (eds.) Homogenization Techniques for Composite media. Lecture Notes in Physics, 272. Springer-Verlag, Berlin (1985)

Whitaker, S.: Simultaneous heat, mass, and momentum transfer in porous media: a theory of drying. Adv. Heat Tran. **13**, 119–203 (1977)

Transp Porous Med (2007) 66:77–87
DOI 10.1007/s11242-006-9023-1

ORIGINAL PAPER

Theoretical models of vegetable drying by convection

Stanisław Pabis

Received: 8 November 2005 / Accepted: 26 March 2006 /
Published online: 24 January 2007
© Springer Science+Business Media B.V. 2007

Abstract The results of experiments are often used to model empirical phenomena. However, the term model is applied in various meanings. A model is usually treated as an abstract formal structure that can replace a material system considered as original, in respect to the aim of modeling. Certain formal structures may be treated as theoretical models of empirical phenomena. On the other hand, a material system can also be referred to as a model of an abstract system, e.g., a set of equations or a hypothesis. Such a material system, if it is a distinct empirical interpretation of the language of a given theory, is then called a real model. Both kinds of models are applied in drying technology, but the second one is more inventive. The mathematical structures are treated as empirical formulae or as theoretical models properly derived from true or legitimated promises of a given theory. The advantages of some mathematical theoretical models of drying processes versus empirical formulae are discussed. The creation of new mathematical theoretical models of convection drying kinetics of some shrinking solids is presented and analyzed. One of the above models was also hypothetically suggested for modeling the drying of cut vegetables in a fluidized-bed. Despite its initial acceptance due to peer empirical justification on cut carrots and celery, it still requires further theoretical analysis. Other models indicated here are theoretical models of vegetable drying in a tunnel drier. These models are created by deduction from laws of heat and mass transfer theory and its basic equations.

Keywords Drying · Vegetables · Mathematical models · Theory · Hypotheses

---

XI Polish Drying Symposium, Poznań, Poland, 13–16 September 2005.

S. Pabis (✉)
Polish Society of Agricultural Engineering,
ul. Polna 54/17, 00-644 Warszawa, Poland
e-mail: spabis@wp.pl

Abbreviations

$A(M)$	Surface area of product with water content $M < M_0$ (m^2)
A_0	Initial surface area of dried product (m^2)
K	Drying coefficient (min^{-1})
L	Latent heat of water vaporisation at product surface temperature (kJ kg^{-1})
M_0	Initial average water content of product (kg kg^{-1})
M_d	Dry matter of product (g)
M_e	Equilibrium water content (kg kg^{-1})
$M(\tau)$	Water content (kg kg^{-1})
N	Exponent in Eq. 11
P_0	Mass of wet ($M = M_0$) product (kg)
P_d	Mass of dry ($M = 0$) product (kg)
V_0	Initial volume ($M = M_0$) of dried product (m^3)
V_d	Volume of dry ($M = 0$) product (m)
$V(M)$	Volume of product with average water content $M < M_0$ (m^3)
b	Value of shrinkage coefficient if $M = 0$, dimensionless
k	Varying rate of drying (kg kg^{-1} min^{-1})
k_0	Constant drying rate (kg kg^{-1} min^{-1})
n	Empirical coefficient
t	Drying air temperature (°C)
t_A	Surface temperature of dried product (°C)
α	Heat transfer coefficient (kJ m^{-2} min^{-1} K^{-1})
ρ_o	Initial density of moist product ($M = M_0$) (kg m^{-3})
ρ_d	Density of dry product ($M = 0$) (kg m^{-3})
τ	Drying time (min)

Mathematical theoretical models of empirical systems (processes being dynamic systems) are a crucial component of science constituting a knowledge of the world. The term model is, however, understood in different ways and therefore it is necessary to determine the way it is understood in the present work. The structural definition of the term model is too complicated and its knowledge is not necessary in our further discussion. At present it is sufficient to provide an intuitive definition of a model as a structure similar to the original. However, this requires a prior definition of the term similar. A precise definition of this term is also equally complicated, but an intuitive understanding of the term that will be presented in a demonstrative way will serve the purpose here. More information on the subject of models and similarity with reference to the processes of drying can be found in "Grain drying: theory and practice" by Pabis et al. (1998). The basic relations between the original and its model are indicated by the charts included in Figs. 1 and 2.

Figure 1 refers to a reproductive situation. A model is to replace the original, being a material system, in order to meet a particular aim. Results obtained from experiments carried out on the already existing original material, e.g. an empirical process, can serve to create its model. A model is an abstract system, e.g. a system of equations. If the model is an empirical formula, it reproduces only the results of measurements carried out on the original. Such a model can also provide information about an action of the original, e.g. a process, but only under such conditions in which measurements of the action of the original were carried out. If the model is a theoretical mathematical

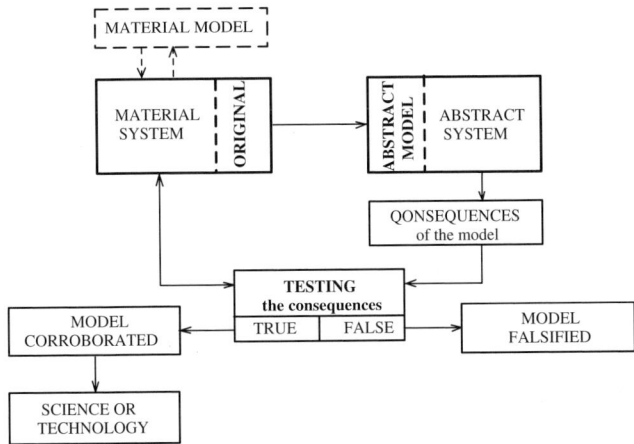

Fig. 1 Forming an abstract system as an abstract model of material (empirical) system

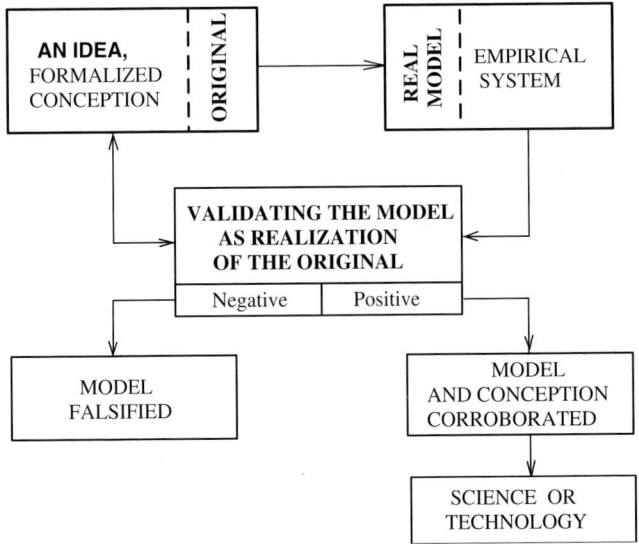

Fig. 2 Material (empirical) system as a real model of a conception, an idea

one, it informs us not only about the action but also about the structure of the original material system. Every model expressed in the language, an abstract model, must be tested empirically.

Figure 2 presents a creative situation.

Human intellectual creativity may yield new ideas related to the development of science or technology. The original is then an abstract (language) system presenting the new idea. This can be, for instance, a hypothesis explaining some phenomenon, a new system of equations which demonstrates or explains something. The truth of such an original, being a language construct but referring to material reality, is tested both directly on the model and indirectly, exploring the truth of consequences resulting from it. The consequences, expressed in the language of mathematics, are checked then in the already-existing material system or one that is being designed, treated here as a model of the idea. Such a way of testing the drying processes is not only creative for science and technology but also the most effective, and thus recommendable. It must be stressed here that a positive result of testing is only a corroboration and not a proof of the truth of the original, e.g. a hypothesis.

The charts presented in Figs. 1 and 2 allow to provide a conventional (conceptual) determination of similarity between the original and its model. Whenever we say that a model is similar to the original we mean that the model may replace the original with respect to the aim of modeling or that the model can test the original on the truth.

Mathematical models of processes of drying solid bodies can therefore be either empirical formulas or theoretical models. Empirical formulas state the existence of a new empirical regularity or can corroborate more precisely the already known regularity, but they do not provide any scientific explanation. They present an insignificant value for science and limited importance for technology. However, they can be very useful for testing theoretical models since they make it possible to reproduce the results of measurements of the action of the studied originals in a precise and continuous way. Comparing results obtained from a theoretical model of the original with the results derived from empirical formulas that are obtained from measurements of action of the original, we determine modeling errors and evaluate models. Prior to the commencement of measurements of the studied process, one should have, at least only a hypothetical concept of its theoretical model, and start to elaborate results of the measurements by means of known models of a given theory. If they do not comply with the theory, this fact may lead to new scientific or technological discoveries.

The way of improving a mathematical model of a theory, demonstrated in Fig. 2, will be presented on an example of a kinetics model for drying cut vegetables by convection. The term kinetics of drying denotes that the model determines the process of drying single solid bodies or their thin layer. A thin layer is a mono-grain layer or a layer of such thickness at which the gradient value of the temperature of the drying gas can be ignored.

The theoretical model here is a model formed deductively from true statements of theory, possibly completed with corroborated hypotheses. The structure of the theoretical model results logically from the theory. Only physical properties of material objects as well as coefficients conditioning probabilistic action of the model are determined empirically in it. The model of drying cut vegetables by convection will be presented here as briefly as possible. Detailed information on its formation and testing can be found in the publications: Pabis (1994, 1999) as well as Pabis and Jaros (2002).

Measurements of drying by convection of single particles of cut vegetables of many kinds almost from the very beginning demonstrate their decreasing rate of drying. Therefore many authors of publications concerning the subject, e.g. Yapar et al. (1990), Mulet et al. (1989a,b), Roman et al. (1979), Vacarezza et al. (1974), have been modeling such processes by exponential equations without any theoretical analysis of the model of drying. Such equations are simplified forms of a theoretical equation in the second period of drying solid bodies with an appropriately low initial water content

$$M(\tau) = M_e + (M_0 - M_e)\exp(-K\tau), \tag{1}$$

where $M(\tau)$ is the average water content of dried product in kg kg^{-1} at the drying time τ in minutes, M_e the equilibrium water content in kg kg^{-1}, M_0 the initial average water content of product in kg kg^{-1}, and K is drying coefficient in min^{-1}.

The equation is an equation of the second period of drying and despite the fact that it well reproduces the kinetics of drying cut vegetables, it has no theoretical rationale in the initial period of drying very moist vegetables. In compliance with the mathematical, classical theory of drying solid bodies by convection, e.g. Newman (1931), Sherwood (1931), Luikov (1950), the very moist cut vegetables should dry in the initial period of drying at a constant rate, according to the equation:

$$M(\tau) = M_0 - k_0 \tau, \tag{2}$$

where k_0 is the constant drying rate in kg kg^{-1} min^{-1}.

In this equation

$$k_0 = \frac{A_0 \alpha}{M_d L}(t - t_A), \tag{3}$$

where A_0 is the initial surface area of dried product in m^2, α heat transfer coefficient in kJ m^{-2} min^{-1} K^{-1}, t the drying air temperature in C, t_A the surface temperature of dried product in °C, M_d the dry matter of product in kg kg^{-1}, L latent heat of water vaporisation at the product surface temperature, and kJ kg^{-1} is constant, under established conditions of the process and it equals the rate of drying. Equation 2 does not prove correct, however, during the drying of cut vegetables with very significant initial water content, sometimes even higher than 10 kg kg^{-1}.

It was then necessary to attempt to elucidate this incompatibility, asking an important general question: does the mathematical classical theory of drying solid bodies by convection really fail to prove correct in the first period of drying cut vegetables?

In consequence, this question has led to the first operational question: why do the results of calculations obtained from the theoretical model presented by equation (2) fail to comply with the results of measurements?

In such a situation, as well as in many similar ones, one can search in different, but methodologically justified ways, for answers providing an explanation, beginning with the general question formulated above. Since methodological foundations of science are formed by mathematical logic, its methods of inference will therefore be used, expressed in the material (semantical) implication of the true statement: "if a solid body dries by convection in the second period of drying, its rate of drying is decreasing", which can be noted in the following formal way:

$$\text{II} \Rightarrow \frac{dM}{d\tau} \neq \text{const} < 0 \tag{4}$$

where the symbol II denotes the second period of drying.

The statement is logically true, which results from the theory, and empirically true, which results from measurements. However, an analysis of the implication indicates that from the fact that for a certain process the rate of drying $dM/d\tau$ is decreasing, it does not follow logically (single-valued) that the process occurs exclusively in the second period. This process can also occur in the first period, despite the fact that it is not indicated by the results of measurements. Results of the logical analysis of this situation not only provide justification for the need but also indicate the necessity to undertake research aimed at elucidating this (apparent) paradox. Every sequence of

measurements of every studied process should be always elucidated theoretically. The development of science as well as technology depends on it.

In the situation when, during the elucidation of new phenomena or empirical regularities, we do not yet know adequate scientific laws, we should begin to interpret these regularities by putting forward explanatory hypotheses, or statements expressing their probable explanation, but that requires testing. In this case, explanation of the incompatibility of measurements of the process with its theory, discussed above, begins with the first hypothesis: *mathematical theoretical models of the kinetics of drying solid bodies by convection, based on the essentially classical theory of drying can be models of the kinetics of drying cut vegetables.* A consequence follows logically from this hypothesis, which is also another hypothesis (explanatory hypothesis), but of a lower degree than the previous one, indicating that *incompatibility of the theoretical linear model (2) with the results of measurements can be caused by not taking account of shrinkage of the dried body during the process of drying.*

This hypothesis is logically corroborated by Eqs. 2 and 3, which directly indicate that when the dried body shrinks, then $A \neq$ const. but $A = A(M)$, thus also $k \neq$ const. but $k = k(A(M))$ =var., which implies that $dM/d\tau \neq$ const. A logical consequence of the second hypothesis is therefore a statement that in the first period of drying, the condition indicating that the rate of drying must be constant need not be met, and therefore that the process must occur in a linear fashion according to Eq. 2.

Thus, we have a few not very difficult problems to solve. We can begin—and we will use this procedure—by forming a non-linear but nevertheless theory-based, new equation of the kinetics of the drying process in its first period. This equation should be tested in a methodologically correct way to examine whether it fulfils the second hypothesis, and at the same time whether it corroborates the first hypothesis. Now we must ask the following question: *how to make the area A of the dried body dependent on its water content M?*

It can be done empirically, yet we should not begin in this way; we should question if it is possible to use the known statements of science for this purpose? There is such a possibility since we know in solid geometry that $A = aV^{2/3}$, where a depends only on the shape of the body, thus there is a theoretically justified empirical possibility that during the process of drying vegetables

$$\frac{A(M)}{A_0} = \left[\frac{V(M)}{V_0}\right]^{2/3}, \qquad (5)$$

where $A(M)$ is the surface area of product with water content $M < M_0$ in m^2, V_0 the initial volume of dried product in m^3, and $V(M)$ is the volume of product with average moisture content $M < M_0$, m^3. This equation is true in mathematics, but it should be tested to see if it is also true in the material world.

Leaving this doubt unanswered for the time being, we make use of the commonly known (e.g. Suzuki et al. 1976; Karathanos et al. 1993; Murakowski 1994) results of measurements of volume shrinkage of dried vegetables, expressed in the equation:

$$\frac{V(M)}{V_0} = aM + b, \qquad (6)$$

where b is the maximum value of shrinkage coefficient, dimensionless, including, unfortunately, two empirical coefficients, a and b. The value b can be easily determined from the results of measurements, however, because when $M = 0$ then $\frac{V(M=0)}{V_0} = b$,

or deducing it (Pabis 1994). Since $V(M = 0) = V_d = P_d/\rho_d$ and $V_0 = P_0/\rho_0$ then after replacing P_0 by $P_d(1 - M_0)$ we obtain:

$$b = \frac{\rho_0}{\rho_d(1 + M_0)}, \qquad (7)$$

where ρ_0 is the initial density of moist product $(M = M_0)$, kgm^{-3}, ρ_d the initial density of dry product $(M = 0)$, kgm^{-3}, P_0 the mass of wet product in kg, and P_d is the mass of dry $(M = 0)$ product in kg.

From Eq. 6 it follows (Pabis 1994), that after mapping M onto M/M_0 the theoretical (and empirically corroborated) model of volume shrinkage of dried cut vegetables can be formed as below:

$$\frac{V(M)}{V_0} = (1 - b)\frac{M}{M_0} + b. \qquad (8)$$

We can now return to Eq. 5, but we must introduce into it an empirical coefficient n to take account of non-isotropic shrinkage of vegetables and shriveling of their surface, thus

$$\frac{A(M)}{A_0} = \left[(1 - b)\frac{M}{M_0} + b\right]^{2/3n}. \qquad (9)$$

Equations 2, 3, and 9 now indicate a non-linear differential equation of the rate of drying of cut vegetables in the initial (first) period of drying:

$$\frac{dM}{d\tau} = k_0 \left[(1 - b)\frac{M}{M_0} + b\right]^{2/3n}. \qquad (10)$$

Solutions of this equation, under the condition when τ equals 0 then M equals M_0, is a non-linear theoretical model of the kinetics of drying cut vegetables by convection:

$$M(\tau) = M_0 \left[\frac{1}{1-b}\left(1 - \frac{1-b}{NM_0}k_0\tau\right)^N - \frac{b}{1-b}\right], \qquad (11)$$

$$N = \frac{3n}{3n - 2} \geq 1.$$

When $N = 1$ then Eq. 11 is transformed into a linear equation (2) of the mathematical (classical) theory of drying solid bodies by convection. This fact is a logical corroboration of the truth of the second hypothesis, and at the same time also the first hypothesis.

Equation 11 was repeatedly and under different conditions of the process of drying a good many times corroborated by measurements of drying by convection of the following e.g.: red beet, Pabis (1999), pumpkin, Sojak (1999), mushrooms, Murakowski (1994). Figure 3 presents comparisons of values calculated from the model (11) with results of measurements.

Equation 11, being a new theoretical model of the kinetics of drying cut vegetables by convection has been empirically corroborated in the range of water content of dried vegetables from about 10 kg/kg to about 2 kg/kg. The maximal error of this model did not exceed locally 5% and the mean value was about 3% under conditions of drying ranging from free convection to the air flow at a velocity of up to 1.2 ms^{-1}.

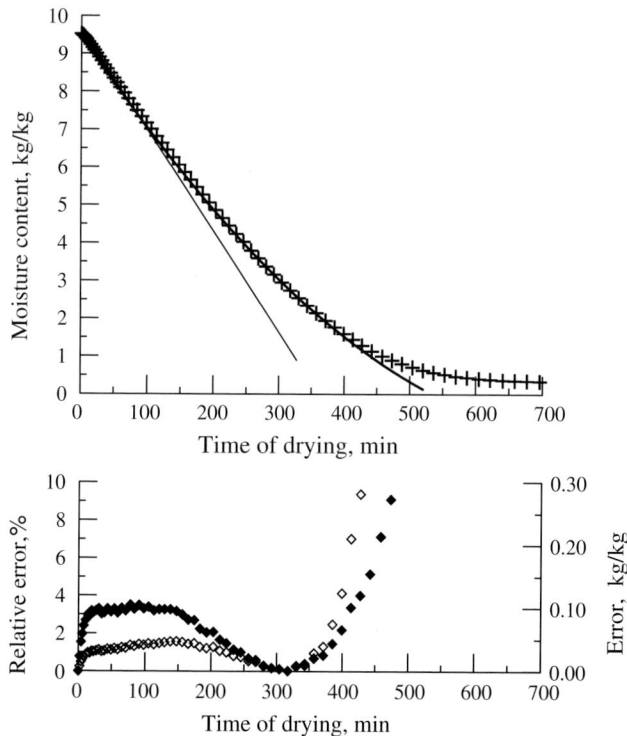

Fig. 3 Mean water content of pumpkin slices (4 × 40 × 50) mm dried at 50°C in free convection and water content predicted by Eq. 11. + measured – calculated if $N = 2$, ♦ error, ◊ relative error

A positive test of Eq. 11 as a new model of the kinetics of drying cut vegetables corroborates thus the second hypothesis being a consequence of the first hypothesis. Thus, we can accept as logical and empirically verified also the first hypothesis that "*mathematical theoretical models of the classical theory of drying solid bodies by convection can be models of the kinetics of drying cut vegetables*".

Having presented this example, the following comments can be made: Eq. 1 in the first period of drying cut vegetables only an empirical formula of the process; Eq. 1 informs in the four-dimensional space, determined by M_0, M_e, K, and τ, about the course of the process in the second period; theoretical equation (11) informs in the 11-dimensional space determined by A_0, α, M_0, M, L, t, t_A, ρ_0, ρ_d, N, and τ, about the course of the process and therefore can provide much more information about it in comparison with the empirical in this period equation (1).

If—and we should aim to achieve this in further research of the processes of drying not only vegetables—tables of values of physical properties of dried products already existed, then by the use of mathematical theoretical models it would be possible to considerably limit natural experiments and expand the experiments simulating the studied processes.

And here are some further examples of theoretical mathematical models of drying vegetables by convection. In the process of fluidization, particles of dried bodies are in a state of considerable loosening in comparison with their state in stationary bed.

It can be hypothetically assumed then that there is hydro-mechanical similarity of the fluidizing particle to the particle suspended freely in the stream of air. Moreover, due to high speeds of drying gas in the fluid bed there, occurs in the bed only a slight temperature gradient of this gas. This situation indicates also similarity to the situation modeled by means of the models of the kinetics of drying cut vegetables discussed here.

These premises inclined Jaros and Pabis (2006) to put forward a hypothesis that *a model for the kinetics of fluidization drying of small-grain or cut solid bodies can be the equation of the kinetics of drying by convection of single particles of such solid bodies, taking into account their shrinkage.* This hypothesis was preliminarily corroborated empirically using results of drying cut carrots and celeries in a laboratory drier. However, this model requires also a theoretical analysis, particularly concerning the transition of the process from the first to the second period of drying. Attention should be paid to the fact that this continuous process is explained by means of different laws of physics, and at the same time different mathematical structures. However, such a situation raises doubts whether this should happen.

A frequently used method for the mathematical formulation of physical issues occurring in process engineering is to start modeling the processes by forming differential equations modeling the given process. Such equations are formulated on the basis of balances resulting from the laws of conservation of mass, energy and motion. In this case, a different situation arises in comparison to the one discussed previously. The method of formulating the model in its differential form is known. Formulating such a model is not necessarily a problem, then, but a task. The problem can be — and often is — an analytical solution of the model posed in its differential form and its transformation to the algebraic form.

The method of balances of mass, energy and motion was used (Markowski and Murakowski 1999) to create theoretical mathematical models of drying cut vegetables in tunnel driers. The mathematically expressed laws of conservation of mass, energy and motion were subjected to a thorough analysis in view of the intention of using them to model drying of cut vegetables. As a result of such an analysis, the assumptions were determined that can be treated as acceptable before beginning to formulate differential equations of this process. A generalized, non-stationary model of the process is described by a system of four differential equations and eleven auxiliary algebraic relations.

After further analysis, the stationary process of drying the mass of cut vegetables, moving in the drier on a trolley was satisfactorily presented using (Piotrowska et al. 1999), four differential equations of the first order and six auxiliary algebraic relations. Despite the fact that this model of the process is theoretically correct, the analytical solution of this system of equations is very difficult. The authors of this model (Markowski et al. 1999) also presented a probabilistic model of the process of drying cut vegetables in a tunnel drier. In this model, the vegetable drying shrinkage was also taken into account.

The stationary model describes the process of co-current or counter-current drying. It was logically tested in experiments carried out on a simulation model. The analysis of the results of computer simulation of this model corroborated its logical correctness. A comparison of the results of the model's simulation with the results obtained from the work of the tunnel drier in the drying plant in Czersk indicated that for the same values of the air temperature at the input to the drier and identical content of water in dried common mushrooms, the final content of water in the mushrooms was almost the same as that obtained from the model. Such compatibility, however, was not indicated by the temperature of the air exiting from the drier and the temperature of dried mushrooms. The results differed by about 5°C with reference to those obtained from the model. At present, we still do not know what causes such a difference between the measurement and the theoretical model.

In relation to the modeling of drying processes, the following general practical comments can be made:

- Studies of vegetable drying processes and elaboration of the results of measurements of the process should always begin with an analysis of the present state of mathematical theory of drying solid bodies.
- Incompatibility between results from measurements and results obtained from theory is also valuable information for science and practice. It may inspire problems of creative scientific or technological research.
- Theoretical mathematical models of drying processes allow also to provided information about results of experiments that have not yet been carried out yet, which is one of the functions of science of crucial importance for technology.
- Practical application of these scientific possibilities, requires however, a systematic gathering and tabulation of information concerning these physical properties of vegetables, which occur in mathematical models of vegetable drying.

References

Jaros, M., Pabis, S.: Theoretical models for fluid-bed drying of cut vegetables. Biosyst. Eng. **93**(1), 45–55 (2006)

Karathanos, V., Anglea, S., Karel, M.: Collapse of structure during drying of celery. Drying Technol. **11**(5), 1005–1023 (1993)

Luikov, A.V.: Theory of Drying (in Russian). Energoizdat, Moscow (1950)

Markowski, M., Murakowski, J.: Formulating the theoretical models of vegetables drying in tunnel driers (in Polish). In: Pabis, S. (ed.) Convection Drying of Vegetables, pp. 69–95. Kraków, Polskie Tow. Inż. Roln.(1999)

Markowski, M., Piotrowska, E., Murakowski, J.: Simplified probabilistic model of vegetables drying kinetics (in Polish). In: Pabis, S. (ed.) Convection Drying of Vegetables, pp. 107–111. Kraków, Polskie Tow. Inż. Roln. (1999)

Mulet, A., Berna, A., Rosello, C.: Drying of carrots. I. Drying models, Drying Technol. **7**(3), 537–557 (1989a)

Mulet, A., Berna, A., Rosello, C. Pinaga, F.: Drying of carrots. II. Evaluation of drying models, Drying Technol. **7**(4), 641–661 (1989b)

Murakowski, J.: Drying of mushrooms in free convection: evaluation of shrinkage, (in Polish). Zeszyty Problemowe Postępów Nauk Rolniczych **417**, 145–154 (1994)

Newman, A.B.: The drying of porous solids: diffusion and surface emission equations. Am. Inst. Chem. Eng. Trans. **27**, 203–220 (1931)

Pabis, S.: Generalized model of drying rate within first period of drying vegetables and fruits (in Polish). ZPPNR **417**, 15–34 (1994)

Pabis, S.: The initial phase of convection drying of vegetables and mushrooms and the effect of shrinkage, J. Agric. Engng. Res. **72**, 187–195 (1999)

Pabis, S., Jaros, M.: The first period of convection drying vegetables and the effect of shape-dependent shrinkage. Biosyst. Engin. **81**(2), 201–211 (2002)

Pabis, S., Jayas, D.S., Cenkowski, S.: Grain Drying: Theory and Practice. John Wiley & Sons, Inc. New York (1998)

Piotrowska, E., Markowski, M., Murakowski, J.: Mathematical model of drying shrinking vegetables in a tunnel drier (in Polish). In: Pabis, S. (ed.) Convection Drying of Vegetables, pp. 97–105. Kraków, Polskie Tow. Inż. Roln. (1999)

Roman, G.N., Rotstein, E., Urbicain, M.J.: Kinetics of water vapor desorption from apples. J. Food Sci. **44**(1), 193–197 (1979)

Sherwood, T.K.: Application of theoretical diffusion equation to the drying of solids. Am. Inst. Chem. Eng. Trans. **27**, 190–202 (1931)

Sojak, M.: Modeling of the kinetics of pumpkin dehydration (in Polish). Inżynieria Rolnicza **2**(8), 87–94 (1999)

Suzuki, K., Kubota, K., Hasegawa, T., Hosaka, H.: Shrinkage in dehydration of root vegetables. J. Food Sci. **41**, 1189–1193 (1976)

Vaccarezza, L.M., Lombardi, J.L., Chirife, J.: Kinetics of moisture movement during air drying of sugar beet root. J. Food Technol. **9**, 317–327 (1974)

Yapar, S., Helvaci, S.S., Peker, S.: Drying behavior of mushroom slices. Drying Technol. **8**(1), 77–99 (1990)

Analysis of the mechanism of counter-current spray drying

Marcin Piatkowski · Ireneusz Zbicinski

Received: 15 October 2005 / Accepted: 26 March 2006 /
Published online: 16 December 2006
©Springer Science+Business Media B.V. 2006

Abstract Results of experimental investigations of the effect of drying and atomization parameters on counter-current spray drying are discussed. Based on 96 experimental tests, the local and global distributions of velocity, temperature, drying air humidity and moisture content of material dried in the drying tower were determined. Analysis of the results showed that the process of agglomeration during counter-current spray drying depended mainly on air temperature in the atomization zone.

Keywords Particle size distribution · Agglomeration · Atomization zone · Discrete and continuous phase recirculation · LDA · PDA

1 Introduction

Spray drying permits a significant extension of the contact area of heat and mass transfer, which can reduce the time of drying to a few seconds and makes this technique applicable to dewatering porous materials, such as many food and pharmaceutical products (Masters 1985).

Spray dryers of different constructions are used in industry. In most cases drying takes place in a concurrent system. Only ca. 5% of spray dryers operate in the counter-current system (Rahse and Dicoi 2001). The process with a counter-current flow of material and drying agent has many advantages, the most important being the reduction of energy input for evaporation and integration of several unit operations in one apparatus, e.g. drying, agglomeration, segregation. However, due to complicated flow hydrodynamics of both phases (recirculation of the continuous phase and agglomeration of particles), drying in the counter-current system is a poorly understood process.

M. Piatkowski · I. Zbicinski (✉)
Department of Heat and Mass Transfer Processes,
Faculty of Process and Environmental Engineering,
Technical University of Lodz, 213 Wolczanska Str.,
90-924 Lodz, Poland
e-mail: zbicinsk@mail.p.lodz.pl

At the Faculty of Process and Environmental Engineering, Lodz Technical University, a drying tower equipped with advanced measuring systems was constructed. These systems permit complex studies on momentum, heat and mass transfer during concurrent and counter-current spray drying (Zbicinski and Piatkowski 2004).

Extensive experimental investigation of spray drying in the counter-current system was performed within this research project.

2 Experimental set-up

A diagram of the experimental set-up is shown in Fig. 1. The main element is a drying tower made of stainless steel, 0.5 m in diameter and 9 m high.

In the experiments a drying agent was hot air aspirated from the atmosphere through a filter. The air was forced into a heater of total power 73.5 kW. The air heated to the required temperature flowed into a distributor (Fig. 2) (Strumillo et al. 2001) equipped with 12 blades. A special mechanism enabled a smooth change of blade position. The slit between the blades could be adjusted in the range 0–10 mm. Depending on the amount and temperature of forced air, inlet tangential velocities ranged from 7 to 57 m/s, which corresponded to mean linear velocities in the column from 0.4 to 0.8 m/s.

Behind the dryer outlet there were 2 cyclones and a bag filter. Purified air was removed to the atmosphere.

Experiments were conducted on a water solution of maltodextrin of initial mass concentration 50%. The solution was prepared in a 150 l tank. The tank was heated with steam supplied to the heating jacket. The solution was supplied from the tank to the atomizer by a mono-pump. It was pumped to a pipeline made from stainless steel and equipped with a heating jacket. A pneumatic nozzle (Spraying System) with a thermostatic water jacket was used to spray the solution.

The spraying nozzle could be located at different distances from the drying air inlet. Tests were made for three distances of the nozzle from the air inlet: 2.4, 4.7 and 6.7 m.

After heating the air to the required inlet temperature and stabilization of thermal conditions in the whole set-up, material spraying was commenced.

Dried (agglomerated) particles of large mass and diameter fell down to the lower cyclone. Small particles that were not agglomerated were entrained by the air and received in the cyclones and bag filter.

During the drying process a number of tests were performed to determine drying kinetics in counter-current spray dryers. The temperature of continuous and disperse phase and air humidity in the dryer were tested on-line. Commercial Dantec systems (Particle Dynamics Analyzer (PDA) and Laser Doppler Anemometry (LDA)), were used to determine the distribution of particle diameters and velocities (Zbicinski and Piatkowski 2004). Samples of material were taken to determine changes in moisture content along the drying path (Delag 2002).

Extremely important drying parameters are the temperatures of material being dried and the drying medium. In the spray dryer, material particles appear as a disperse phase which is in close interaction with the continuous phase. The use of classical thermocouples allows us to measure the approximate temperature of the mixture of particles and drying air. The approximation is a result of the fact that particles moving in an air stream collide with the thermocouple and deposit on it.

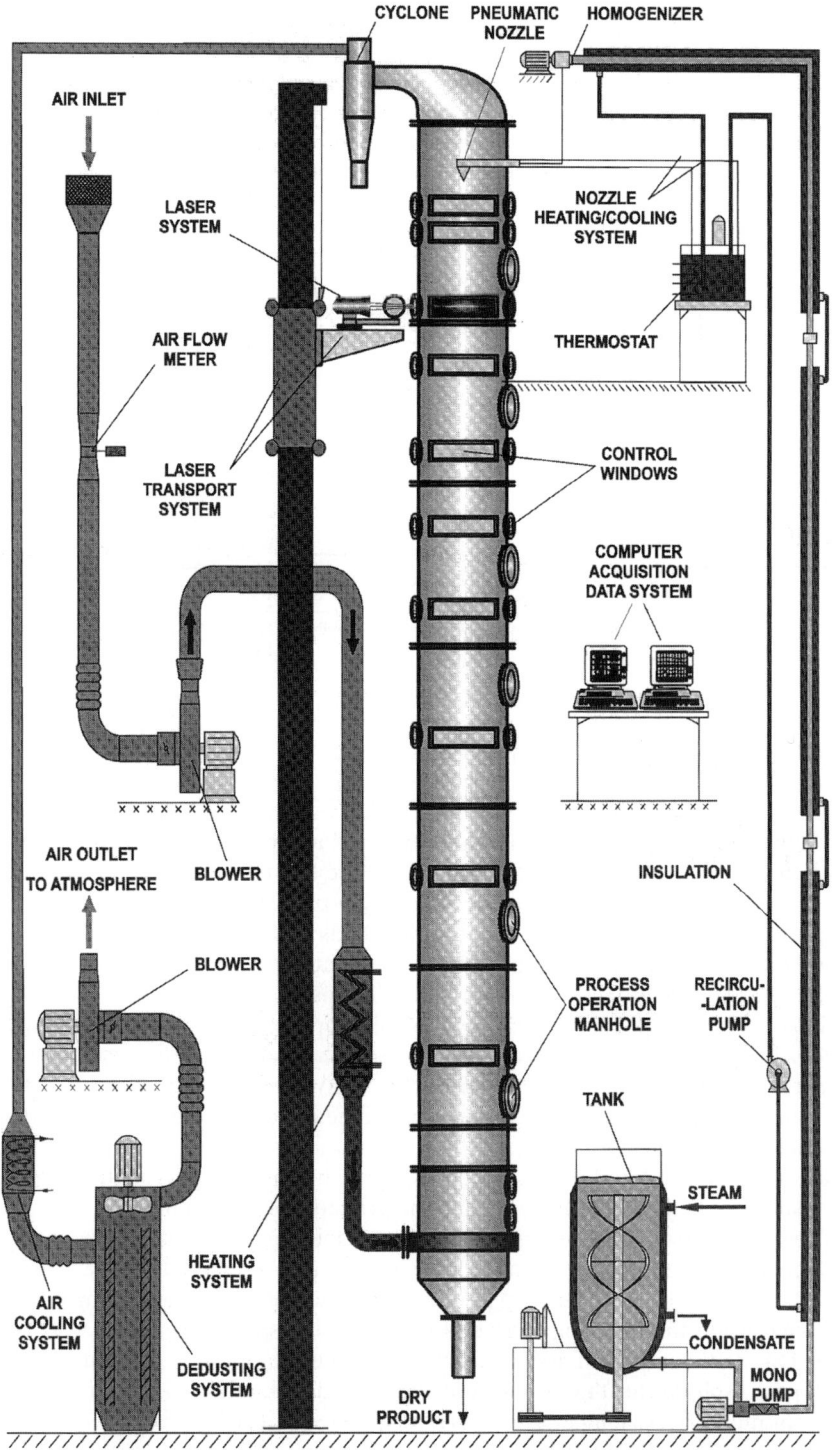

Fig. 1 Experimental set-up

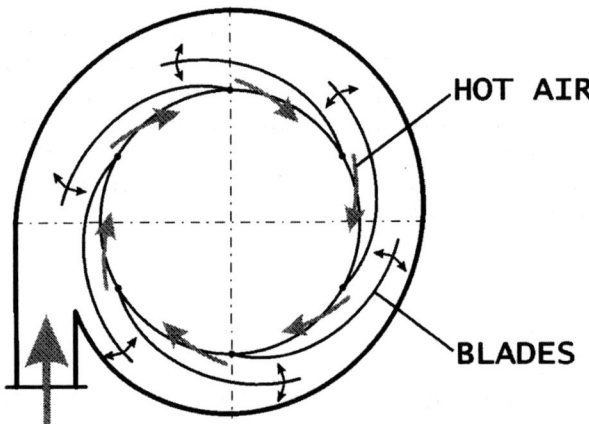

Fig. 2 Schematic of air inlet

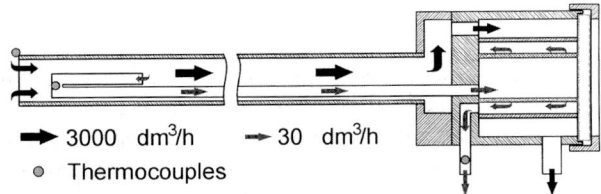

Fig. 3 Schematic of the microseparator

Many methods were tried to eliminate errors that appear as a consequence of this phenomenon. The most frequently applied method was to use a shielded thermocouple. A partly shielded thermoelement enables determination of the continuous phase temperature with satisfactory accuracy (Papadakis and King 1989; Zbicinski 1995). However this method cannot be applied in the counter-current drying system because dried material particles move in all directions in this process.

In our tests we applied a dynamic method of temperature measurement in the continuous phase, in which inertia forces were used to separate the disperse phase. The idea of this method, called "microseparator", was given by Kievet and Kerkhof (1996). The name of the method is related to the phenomenon of aerodynamic separation of particles from the air.

The solution proposed by Kievet and Kerkhof has been improved in our project. Fig. 3 shows a diagram of the device and the principle of operation.

The outer pipe of greater diameter is used to aspirate the mixture of drying air and dried particles from the dryer. In the pipe, the air and particles flow at a significant rate (ca. 3.5 m/s) in the pipe. A pipe of smaller diameter, bent at an angle of 180° is placed inside the outer pipe. This pipe is used to aspirate the air at velocity ca. 1 m/s. Due to flow rate difference in both pipes and the opposite flow direction, the small pipe aspirates air that is almost completely devoid of dried material particles. The continuous phase temperature was measured using a thermocouple placed inside the small pipe. Additionally, a thermocouple was installed at the end of the larger diameter pipe to measure the jet temperature.

Separation of the continuous and disperse phase allowed us to measure one more important process parameter—moisture content of the continuous phase.

A precision S4000 Michell dew-point thermometer (accuracy 0.01 °C) was used for air humidity measurements. After the particles were removed, the air was supplied to a hygrometer by a heated Teflon pipe 5 m long. Construction of the microseparator enabled free displacement of the device and the performance of tests on 12 measuring levels. On each level the temperature and humidity were determined at 7 points along the dryer radius.

3 Scope of investigations

The full scope of the investigations covered many aspects of spray drying kinetics in counter-current systems. In this paper we focus on the effect of process parameters on the degree of particle agglomeration in the drying tower.

4 Analysis of results

A laser anemometer (LDA) was used to measure local velocity distributions of dried material particles. Disperse phase velocity was determined in the tower axis and at 6 points located on the dryer radius. Fig. 4 shows the directions of particles in the dryer.

Particles and droplets of dried material were displaced with drying air axially, tangentially and radially. Lighter particles were entrained upwards to the atomization zone where they encountered the stream of sprayed material. As a result of contact between particles (droplets), agglomerates were formed, which dropped down the column and contacted the drying agent of growing temperature. Mass loss due to evaporation might cause the agglomerate to be raised again by the drying air to the atomization zone where it was further agglomerated.

The process of agglomeration proceeded until the moment when the force of gravity on the agglomerated particle was greater than the forces of aerodynamic resistance of the drying agent. The particle then fell below the air inlet to the lower receiver.

Diameters of dried material particles were measured by a PDA laser measuring system below the atomizer and one level above it, nearest to the air outlet from the dryer.

The laser system provided information on the diameters of several thousand particles from each measuring point. The data were sorted and divided into ranges, every 10 μm. On the basis of the number of particles in particular fractions, their percentage was calculated.

Figure 5a–f show local particle size distribution in such drying conditions in which final particle diameters were the biggest. The applied PDA configuration enabled data acquisition for particles of diameters up to 1000 μm. Since the fractions of particles with diameters greater than 700 μm did not exceed 0.2%, the range of diameters shown on the graphs was limited to 0–700 μm. Analysis of Fig. 5a leads to the conclusion that the diameter distributions are most unified above the atomizer. In this part of the column, there were particles entrained with the drying air whose diameters were relatively small with moisture content not exceeding 5%.

The diagram shown in Fig. 5b (distance from the air inlet 4.6 m) shows particle size distribution at a distance of 0.1 m from the atomizer. In the dryer axis and at a

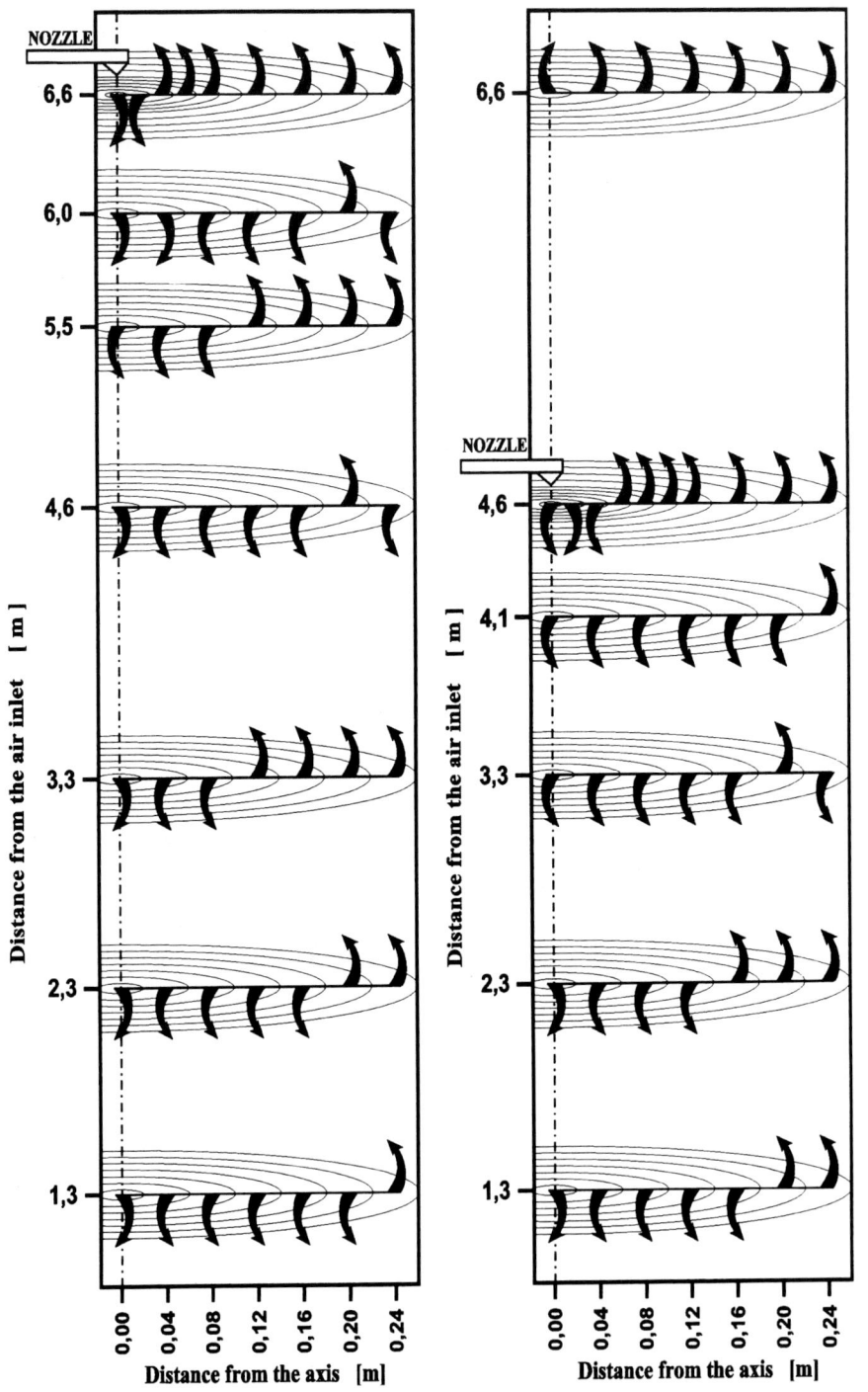

Fig. 4 Directions of particles in the dryer (air 300 Nm^3/h, temperature 150°C, slit between blades 10 mm, solution 6 kg/h, atomizing air 1.8 kg/h, nozzle 4.7 and 6.7 m)

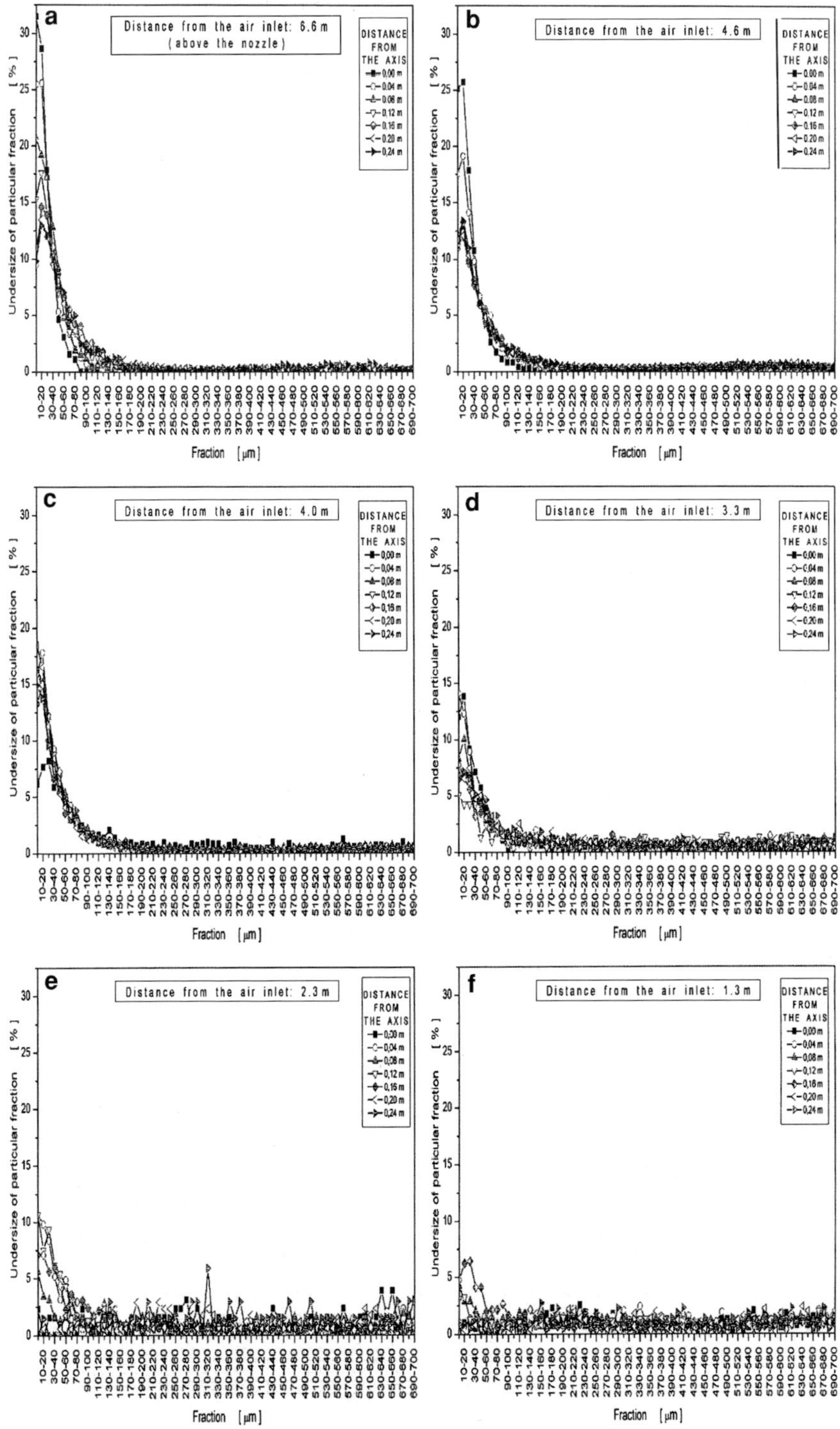

Fig. 5 Fractions of particles (air 300 Nm^3/h, 150°C; slit between blades 10 mm, material 6 kg/h, atomizing air 1.8 kg/h, distance of the nozzle from the inlet 4.7 m)

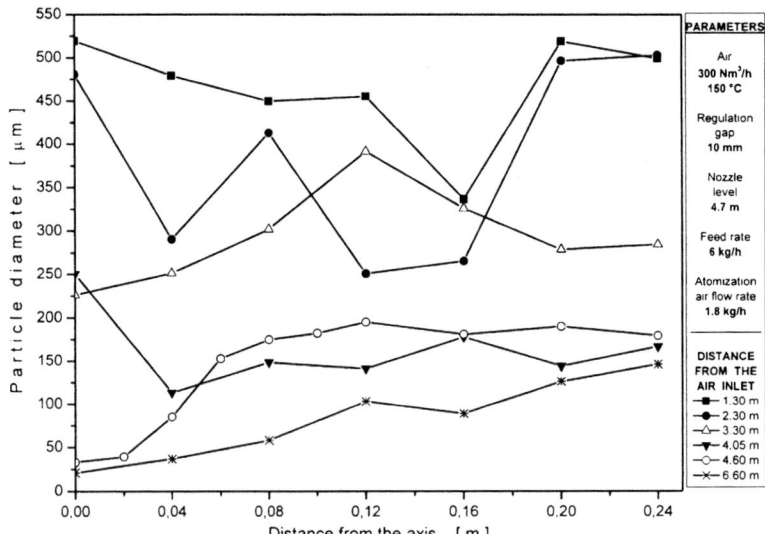

Fig. 6 Cross-sectional distributions of particle diameters

distance 0.04 m from the axis, where a compact jet of sprayed material is observed, the percentage of particle fractions with the smallest diameters reached 26%. At distance 0.24 m from the axis, the percentage of the smallest fractions did not exceed 15% and particles of diameters exceeding 400 μm appeared at the same time; this is caused by recirculation of the drying air stream and entrainment of single particle agglomerates.

On analyzing Fig. 5c–f, one can observe significant changes in particle diameter distribution during drying. With increasing distance from the atomizer, the percentage of particles with diameters exceeding 150 μm increases and those with diameters up to 50 μm decreases. There was a uniform distribution of particle diameters at a distance 1.3 m from the air inlet.

In order to better illustrate the above results, Fig. 6 shows mean arithmetic particle diameters at successive measuring points.

In Fig. 6 one can observe changes of particle diameters in the drying tower cross section. Above the atomizer (6.6 m from the air inlet) particle diameters are distributed along the column radius, from the smallest one in the column axis to the biggest near the wall. This distribution is a result of entrainment of the particles from the atomization zone (small particles in the axis) and aerodynamic segregation (bigger particles near the wall).

It can also be observed that at distance 0.1 m below the atomizer (4.6 m from the air inlet) at a width 0.04 m from the axis the sprayed material stream was still flowing. Particles entrained by the air flowing from the lower part of the column appeared further from the axis.

Subsequent curves show how particle diameter changed with increasing distance from the atomizer. Near the air inlet the mean particle diameter was the greatest, reaching ca. 520 μm.

Irregular changes of particle diameters along the radius at the distance smaller than 3.3 m from the air inlet (Fig. 6) are a result of particle recirculation in the column.

Figures 7–9 show changes of particle diameters as a function of distance from the air inlet for 24 drying conditions. It was found that the final agglomerate size depended on process condition and changed over a broad range.

Analysis of Figs. 7–9 shows that the biggest influence on particle diameters was the atomizer position. When it was changed, the temperature of air in direct contact with atomized particles also changed. This means that the process of agglomeration takes place mainly in the atomization zone, i.e. in the area 1–2 m below the atomizer.

It follows from analysis of Fig. 7 that higher gas temperature in the atomization zone hampered agglomeration and final particle diameters in these conditions did not exceed 150 μm in most cases. Poor agglomeration of particles was a result of rapid moisture evaporation and a lack of sufficient number of wet particles that were most susceptible to agglomeration.

Displacement of the atomizer from the distance of 2.4 to 4.7 m from the air inlet (Fig. 8) caused a decrease of air temperature in the atomization zone by about 10°C (Fig. 10). In these conditions the agglomeration process was efficient and mean particle diameters varied from 250 to 450 μm.

Location of the atomizer 6.7 m from the air inlet caused a further gas temperature drop in the atomization zone (Fig. 10). A consequence of the temperature drop in the agglomeration zone was a change of product particle diameter distribution. Agglomerates obtained in these conditions had mean diameters up to 450 μm; the smallest particle diameters obtained in this location of the atomizer did not exceed 150 μm (Fig. 9).

Figure 11a–f show the structure of particles obtained in different positions of the atomizer. The photographs presented were taken with a Hitachi S-3000N electron microscope. Photographs in the left column show material from the bottom collector, while the right one shows photographs of powder separated in the cyclones. Magnification in the right and left column differs five times.

In the present drying conditions, particles obtained from the cyclones had similar shapes and diameters. These were non-agglomerated particles of diameters equal to dozens of micrometers.

Powders collected in the lower receiver were agglomerates and as such had a greatly differentiated external structure. Drying processes carried out with the atomizer installed at the level of 2.4 m caused formation of agglomerates which had the structure of combined spheres. This was a result of the previously described intensive mass transfer induced by high temperature in the agglomeration zone.

Particles obtained during spraying from the atomizer located 6.7 m from the inlet can be characterized as particles of uniform structure and large diameters. Lower gas temperature at the atomizer level caused the particles to form large homogeneous structures.

As was indicated earlier, location of the atomizer on the intermediate level, i.e. at a distance of 4.7 m, resulted in the formation of agglomerates with the greatest diameters. Figure 11c shows material particles obtained for the position of the atomizer. Their structure reveals that during drying moisture was partly removed from the particles that then got into contact, which led to the formation of extensive spatial structures.

Summarizing the results obtained, we can conclude that optimum agglomeration conditions can be determined for every process parameter of counter-current drying.

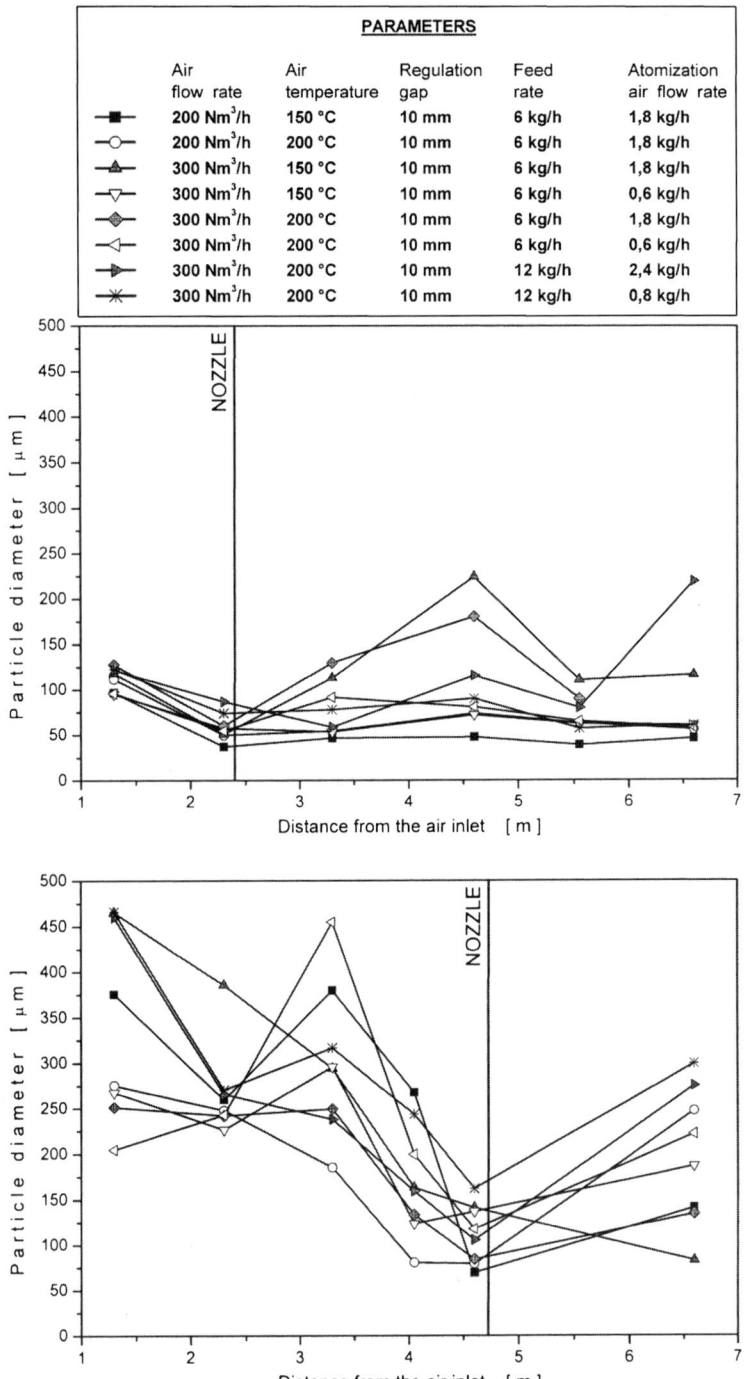

Figs. 7 and 8 Particle diameter distribution along the drying tower

Analysis of the mechanism of counter-current spray drying

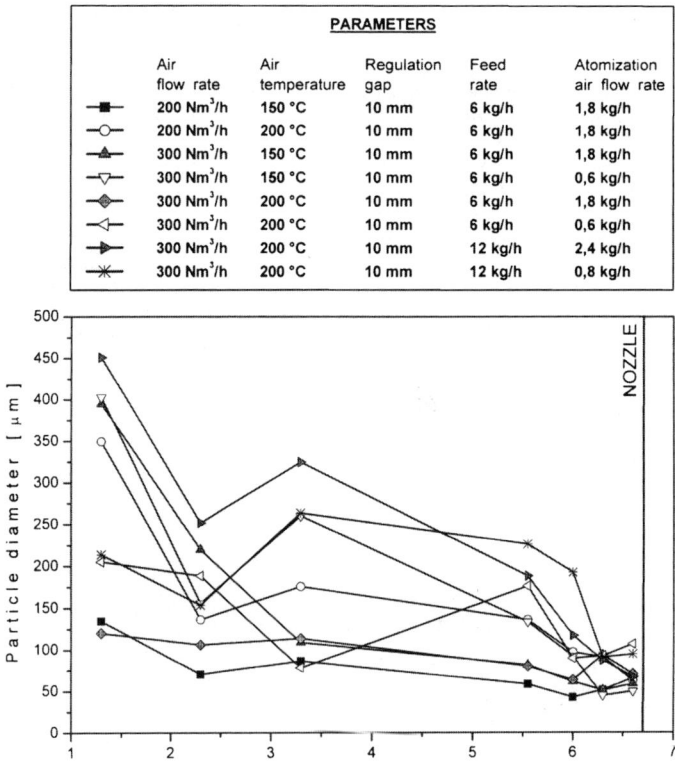

Fig. 9 Particle diameter distribution along the drying tower

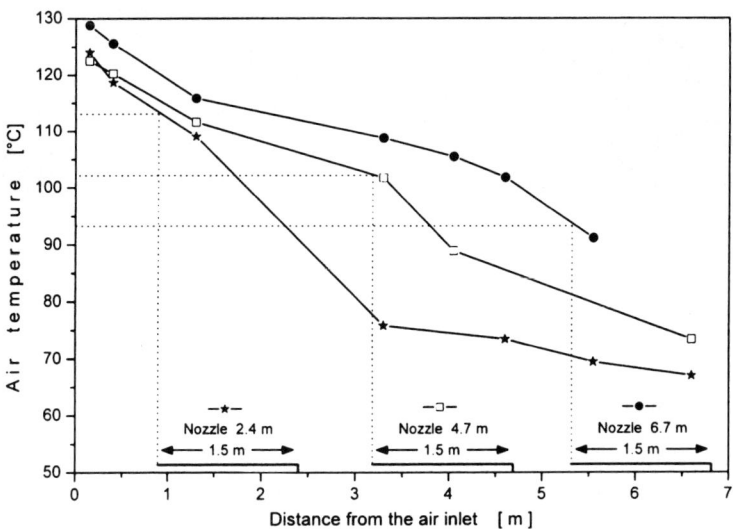

Fig. 10 Drying air temperature (Drying air 300 Nm^3/h, Temperature 150°C, Slit between blades 10 mm, Material 6 kg/h, Atomizing air 1.8 kg/h)

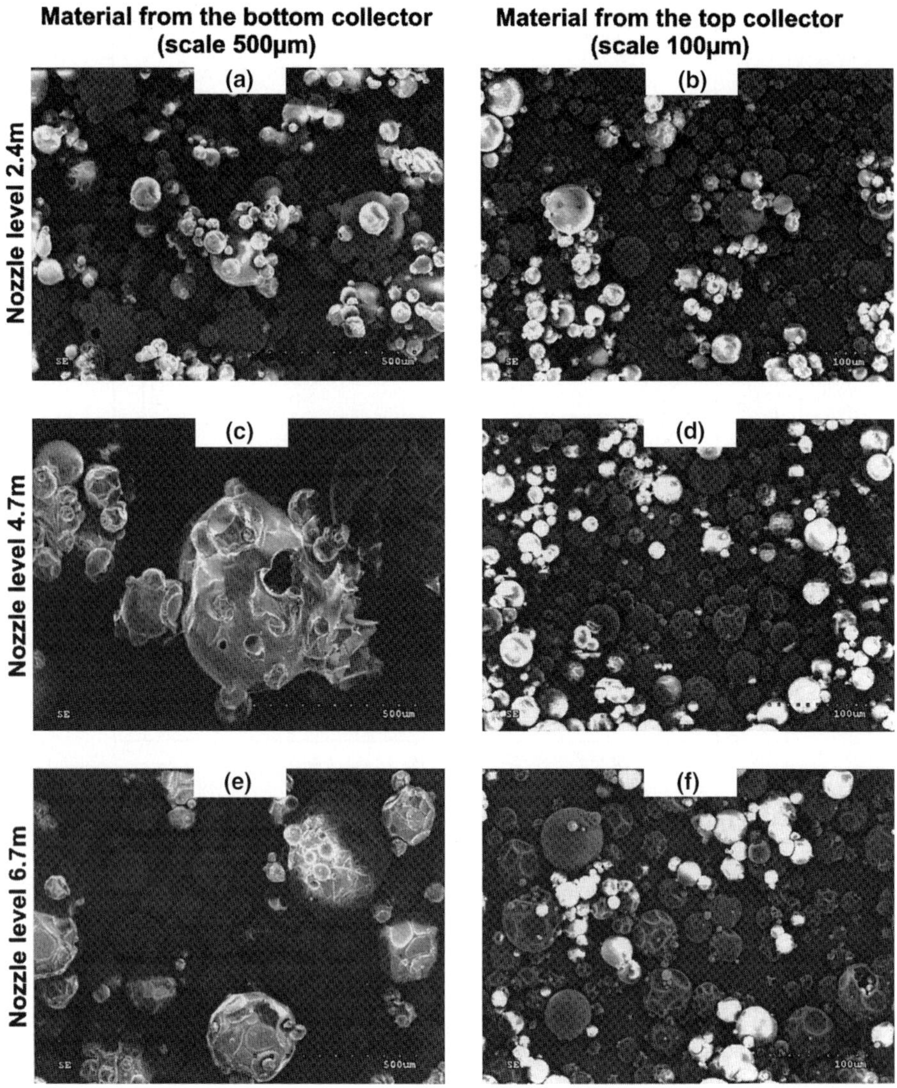

Material from the bottom collector (scale 500µm)　　**Material from the top collector (scale 100µm)**

Nozzle level 2.4m — (a) / (b)
Nozzle level 4.7m — (c) / (d)
Nozzle level 6.7m — (e) / (f)

Fig. 11 Photographs of dried material particles

5 Conclusions

1. Agglomeration during counter-current spray drying depends on drying air temperature in the atomization zone.
2. For every atomization parameter optimum drying conditions, i.e. such in which agglomeration process is most efficient, can be determined.
3. By proper choice of drying and atomization parameters we can control diameters of the agglomerates obtained in counter-current spray drying.

Acknowledgements Experimental work was partly financed by the International Fine Particle Research Institute (USA).

References

Delag, A.: Drying and degradation kinetics in a disperse system. PhD thesis, Lodz Technical University, Poland (2002)

Kievet, F.G., Kerkhof, J.A.M.: Modelling and simulation of drying. In: Strumillo, C., Pakowski, Z. (eds.) Using Computational Fluid Dynamics to Model Product Quality in Spray Drying: Air Flow, Temperature and Humidity Patterns, Drying'96, vol. A, pp. 259–266. Krakow, Poland (1996)

Masters, K.: Spray drying handbook, 4th edn. George Godwin, London (1985)

Papadakis, S.E., King, C.J.: Spray drying and drops. In: Mujumdar, A.S., Roques, M. (eds.) Factors Governing Temperature and Humidity Fields in Spray Drying, Drying'89, vol. 1, pp. 345–352. Versailles, France (1989)

Rahse, W., Dicoi, O.: Spray drying in detergent industry. In: Proceedings of Spray Drying Conference '01. Dortmund, Germany, 2001, 11 (2001)

Strumillo, C., Zbicinski, I., Delag, A., Kwapinska, M., Piatkowski, M., Li, Xuanyou: Scaling-up and predictions of final product properties in spray drying process. Report ARR 35-04 (Florida) for International Fine Particle Research Institute (2001)

Zbicinski, I.: Development and experimental verification of momentum, heat and mass transfer model in spray drying, Chem. Eng. J. **58**, 123–133 (1995)

Zbicinski, I., Delag, A., Strumillo, C., Adamiec J.: Advanced experimental analysis of drying kinetics in spray drying. Chem. Eng. J. **86**, 207–216 (2002)

Zbicinski, I., Piatkowski, M.: Spray drying tower experiments. Drying Technol. **22**(6), 1325–1350 (2004)

Modern modelling methods in drying

Thomas Metzger · Marzena Kwapinska · Mirko Peglow · Gabriela Saage · Evangelos Tsotsas

Received: 15 November 2005 / Accepted: 26 March 2006 /
Published online: 30 August 2006
© Springer Science+Business Media B.V. 2006

Abstract Several modern modelling techniques are presented as tools for drying science and technology, namely pore networks, discrete element method and population balances. After first presenting results from their own research, the authors indicate what future contributions to a better understanding of the drying process at different levels—single porous particles, agitated and fluidised beds—may be expected.

Keywords Drying kinetics · Agitated bed · Mixing · Fluidised bed · Agglomeration · Pore networks · Discrete element method · Population balances

Nomenclature

A	Pore throat cross section [m^2]
c_p	Heat capacity [kJ kg^{-1} K^{-1}]
f	Number density [m^{-3} kg^{-1}]
G	Rate of drying/wetting [kg s^{-1}]
L	Capillary or pore throat length [m]
l	Amount of liquid [kg]
$\dot{M}$	Mass flow rate [kg s^{-1}]
$\tilde{M}$	Molar mass [kg kmol^{-1}]
m_l	Liquid mass density [kg^{-3}]
N_{mix}	Mixing number [–]
n	Rotational frequency [s^{-1}]
n	Number density [s^{-1}] or [m^{-3}]
$\dot{n}$	Number flow rate [s^{-2}]

T. Metzger (✉) · M. Kwapinska · M. Peglow · G. Saage · E. Tsotsas
Chair of Thermal Process Engineering,
Otto-von-Guericke-University,
P.O. 4120, D-39016 Magdeburg, Germany
e-mail: thomas.metzger@vst.uni-magdeburg.de

$\tilde{R}$ Ideal gas constant [kJ kmol^{-1} K^{-1}]
p Pressure [Pa]
r Radius [m]
r_0 Mean radius [m]
s Meniscus position [m]
T Absolute temperature [K]
t Time [s]
t_{mix} Revolution time [s]
t_R Mixing time [s]
v Particle volume [m^3]
z Space coordinate [m]

Greek symbols
α Heat transfer coefficient [W m^{-2} K^{-1}]
β Mass transfer coefficient [m s^{-1}]
β Agglomeration kernel [s^{-1}]
δ Binary diffusion coefficient [m^2 s^{-1}]
ζ Dimensionless position [–]
η Dynamic viscosity [Pa s]
λ Thermal conductivity [Wm^{-1} K^{-1}]
$\dot{\nu}$ Dimensionless drying rate [–]
ρ Density [kg m^{-3}]
σ Surface tension [N m^{-1}]
σ_0 Radius standard deviation [m]
τ Age of particle [s]
ψ Porosity [–]

Subscripts
bed   (Penetration into) Bed
I   First drying period
v   Vapour
ws   Wall-to-bed

1 Introduction

In recent years, new approaches have been taken to the modelling of drying; unlike traditional models that describe the drying process in terms of average quantities and at a continuous level, they are able to give values of local or distributed character. These advances have become possible due to the development of fast computers, which can handle large amounts of data, and new numerical techniques.

In the present paper the following approaches will be discussed: first, network modelling, which is able to describe drying of porous media at the pore level and will permit a systematic investigation of the influence of structure on drying kinetics. For realistic results, simulations with large pore networks are required, which became possible only with increasing computer performance. In the past, intra-particle transport processes were modelled in continuous approaches, such as purely diffusional models, using lumped parameters. In contrast, the new discrete approach aims at a microscopic description of all relevant physical phenomena.

Second, discrete element modelling will be presented, which describes individual particle motion in agitated beds based on mechanical principles. Recent model upgrades include thermal contacts between the particles, so that modelling of heat transfer in contact dryers is possible. The individual description allows us to assess traditional models which use a continuum approach. A full description of the drying process in contact dryers is expected in the near future. As in the case of networks, this kind of modelling only became possible with powerful computers.

Finally, population balances are discussed. These are used in fluidised bed operations to model the evolution of properties that may differ from particle to particle, such as moisture content or temperature. Instead of computing only averages for all particles, as in traditional models, distributions of these quantities are obtained. This permits a better assessment of product quality, even more so as the history of particles and its influence on other important properties, e.g. bioactivity, are accessible.

2 Network models: influence of pore structure on drying kinetics

It is well known that drying kinetics of a porous body depend on two different groups of parameters: one refers to the drying conditions, the other to the material itself. In convection drying, the first drying period is described by temperature, humidity and velocity of the drying air as well as the outer geometry of the body (needed to compute transfer coefficients); the second drying period depends, from its beginning, on the internal structure of the solid. Naturally, the factors determining the constant drying rate are easily accessible whereas the modelling of the second drying period must be based on a range of parameters that need to be measured in well-designed experiments. These parameters can either be effective transport parameters for a partially saturated porous medium that are to be used in a volume-averaged drying model (Perré and Turner 1999); or they may be structural parameters describing pore size distribution, connectivity of pores, etc.

In order to investigate the influence of structural parameters on drying kinetics there are, in principle, two possibilities: one can aim to describe their influence on the effective transport parameters; or one can use the structural information to simulate the drying process directly at the pore level. For both approaches, a suitable tool is pore networks consisting of regularly or randomly located pores that are interconnected by throats. To be representative of a real porous medium, such networks should be chosen large enough and have the same structural properties. At present, these requirements can only be partially fulfilled, because experimental tools to analyse porous media in terms of their structure down to the nanoscale are not available. Therefore, network simulations are qualitative in nature or must be restricted to macroporous structures.

In the past decade, pore network models, which were traditionally used to model drainage, have become very popular in the field of drying (e.g. Nowicki et al. 1992; Laurindo and Prat 1998; Plourde and Prat 2003; Yiotis et al. 2001, 2005; Segura and Toledo 2005). To our knowledge, all networks used for drying simulations were regular ones: square for two-dimensional and cubic for three-dimensional modelling. However, literature models vary in the geometry of pores and throats: throats have cylindrical, rectangular or polygonal cross sections or they are bi-conical to better represent the converging–diverging character of a real pore space; pores are chosen as nodes without volume or as pore bodies. Viscous effects are either completely neglected or accounted for in the liquid phase (Nowicki et al.) or in both liquid and gas phase (Yiotis et al.). Most approaches assume isothermal conditions; heat transfer has only been modelled by Prat and coworkers who also used experiments with micromodels to validate their model. Film and corner flow in gas-filled throats has remained an important issue over the years. Gravitational effects on drying rates have been investigated as well.

In the early research, Nowicki et al. derived, among other properties, effective vapour diffusivity and relative liquid permeability as a function of saturation – using local vapour or liquid fluxes and corresponding gradients in vapour or capillary pressure, as they occur during the drying of the network. However, the influence of network structure on drying behaviour or on effective transport parameters was not investigated. For a long period, other researchers only simulated the drying process for networks – without computing effective transport parameters, and also without variation of the pore structure. Only recently, Segura and Toledo (2005) computed drying curves as well as diffusivity and permeablities for pore saturations of obtained from drying simulations. In their work, pore size distribution as well as throat shape were varied to study structural influences, but the small range of variations in pore size and the absence of viscosity in the model make it difficult to draw conclusions.

On the way to a better understanding of the influence of pore size distribution on drying behaviour, we lately introduced a one-dimensional capillary model (Metzger and Tsotsas 2005; Metzger et al. 2005). The pore space is represented by a bundle of cylindrical capillaries, set perpendicular to the product surface and connected without any lateral resistances (see Fig. 1). Isothermal drying was modelled for a viscous liquid, both for plate and spherical porous objects. The major condition for liquid flow is that between any two menisci the difference in capillary pressure $\Delta p_{c,i}$ must overcome the pressure drop due to friction $\Delta p_{f,i}$. For plate geometry, we have

$$2\sigma \left(\frac{1}{r_{i+1}} - \frac{1}{r_i} \right) = \Delta p_{c,i} \geq \Delta p_{f,i} = \frac{8\eta \dot{M}_v (s_i - s_{i+1})}{\rho \sum_j \pi r_j^4} \quad (1)$$

where σ is surface tension, ρ density and η dynamic viscosity for liquid water, $\dot{M}_v$ total evaporation rate, and r_i are radii of capillaries and s_i positions of respective menisci. Assuming ideal lateral transfer, $\dot{M}_v$ depends only on the mass transfer coefficient β for the boundary layer until the last meniscus withdraws from the surface; then an additional resistance for vapour diffusion appears over the distance s_n.

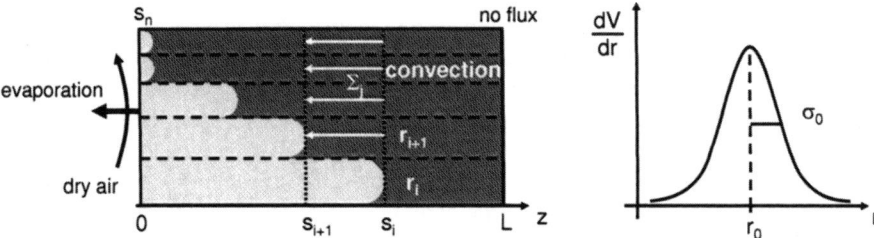

Fig. 1 Bundle of capillaries with normal pore volume distribution

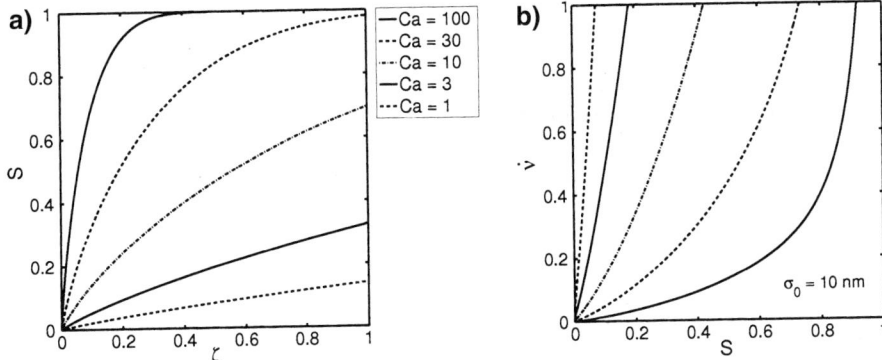

Fig. 2 Moisture profiles at the end of the first drying period and dimensionless drying curves for different Capillary number

For mono-modal pore volume distributions (normal with mean r_0 and standard deviation σ_0) we found that the absolute size of pores plays only a minor role, especially if $\sigma_0 \ll r_0$, but that the duration of the first drying period is significantly increased as the distribution becomes broader. In Fig. 2, saturation profiles for the dimensionless coordinate $\zeta = z/L$ as well as dimensionless drying rates $\dot{v} = \dot{M}_v/\dot{M}_{v,I}$ are plotted for different values of the Capillary number

$$\mathrm{Ca} = \frac{\beta \eta L}{1000 \cdot \psi \sigma \sigma_0} \tag{2}$$

where ψ is porosity (equal to unity if solid phase is not included) and L capillary length. (The factor is to account for the density difference between liquid water and vapour.) For $\beta = 0.1\,\mathrm{m/s}$, $L = 0.1\,\mathrm{m}$, $\psi = 0.5$ and $\sigma_0 = 10\,\mathrm{nm}$, we have $\mathrm{Ca} \approx 30$. Whereas the first drying period only depends on Ca, the second period needs one more parameter, i.e. distribution width σ_0.

For bi-modal distributions drying is generally faster because the large pores dry out first, more or less completely; consequently, their volume fraction significantly influences the drying curve. It is clear that the validity of the model is limited due to its restriction to one dimension and the resulting continuity of the liquid phase. Currently, the geometry of this model is used to derive all relevant macroscopic properties of the corresponding porous medium; these are then employed in volume-averaged drying simulations to show the equivalence of continuous and discrete approach. In this frame, we already extended the capillary model to simulate drying with heat transfer for two (classes of) capillaries of different radius in plate geometry (Metzger and Tsotsas 2005).

In parallel, the isothermal pore network model of Prat was used to study how drying kinetics depend on pore structure (Irawan et al. 2005; Metzger et al. 2005). The network consists of pore nodes connected by cylindrical throats of random radii, which are chosen according to a pore size distribution. Initially, all throats are filled by liquid, which can then evaporate through one surface of the two- or three-dimensional network. During drying, the liquid phase typically becomes discontinuous, splitting up into many clusters. The major model assumption (valid only for relatively large pores) is that capillary forces dominate over friction, so that within one such cluster the throats empty one by one with decreasing radius because liquid is always pumped

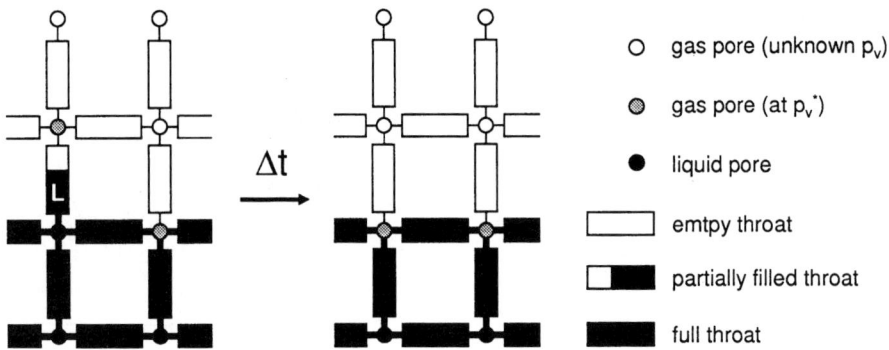

Fig. 3 Emptying of the largest throat (*L*) in a liquid cluster

to small throats. Figure 3 shows schematically the state of pores and throats, before and after the emptying of a throat. The drying rate is determined by vapour diffusion as computed from mass balances for all gas filled pores (*i* and *j*)

$$\sum_j \dot{M}_{v,ij} = \sum_j A_{ij} \frac{\delta}{L} \frac{p\tilde{M}_v}{\tilde{R}T} \cdot \ln\left(\frac{p - p_{v,i}}{p - p_{v,j}}\right) = 0 \tag{3}$$

where A_{ij} is cross section and L length of connecting empty throats, δ vapour diffusion coefficient, p total pressure, p_v vapour pressure, $\tilde{M}_v$ molar mass of vapour, $\tilde{R}$ ideal gas constant and T absolute temperature. We combined the idea of a discretised boundary layer proposed by Laurindo and Prat (1998) with laminar boundary layer theory, so that the network of nodes is naturally extended up to the average boundary layer thickness (see Fig. 4). The boundary conditions for Eq. (3) are then given by the vapour pressure of bulk drying air and saturation vapour pressure p_v^* at any liquid surface inside the network. The amount of water in the largest throat divided by the total evaporation rate for the cluster (sum over all menisci) gives the time step Δt. Emptying of a throat may create new clusters.

Figure 4 shows a typical phase distribution with several liquid clusters along with the vapour pressure field. In the network, vapour pressure is uniform except near the surface so that some of the clusters are not (yet) evaporating; in the boundary layer one can see the lateral vapour transfer.

With this model, we computed drying curves for two- and three-dimensional networks with different structures and pore size distributions; two examples are shown in Figs. 5 and 6; the corresponding moisture profiles are plotted in Fig. 7. Additionally, the influence of boundary layer thickness (or drying air velocity) and product depth were analysed for different structures. It could be shown that the first drying period can only be reproduced if a discretised boundary layer is included in the model.

As a consequence of neglecting any viscous forces, the only counter mechanism to capillary pumping is the discontinuity of the liquid phase: the network dries out rapidly only if capillary flow paths can be maintained over a significant time. For a network 1 with normally distributed throat radii, this is not the case; in the present model, a change of mean pore size and broadness of the distribution will not change the drying behaviour, since the splitting up into clusters is determined by the order of throat emptying which proceeds according to relative decrease in size. For bimodal

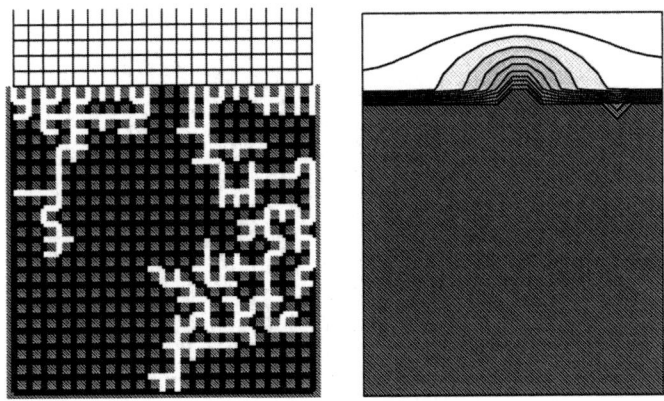

Fig. 4 Pore saturation and vapour pressure field for a small two-dimensional network

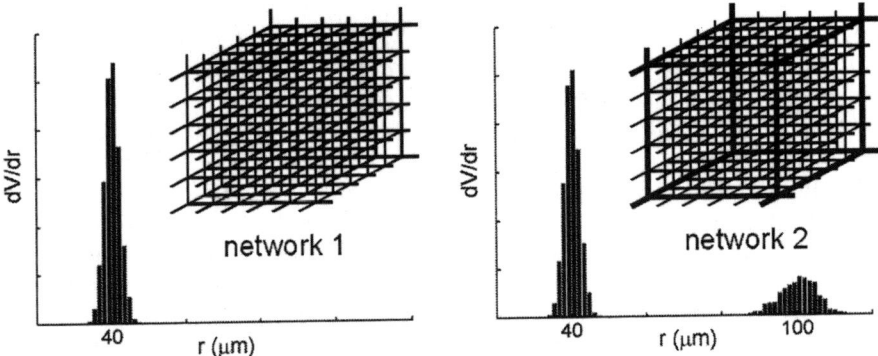

Fig. 5 Pore size distribution and pore structure for two different $16 \times 16 \times 16$ networks with $L = 500\,\mu\mathrm{m}$ (only part of the network is sketched)

Fig. 6 Drying curves for the two networks in Fig. 5 (evaporation flux as a function of network saturation)

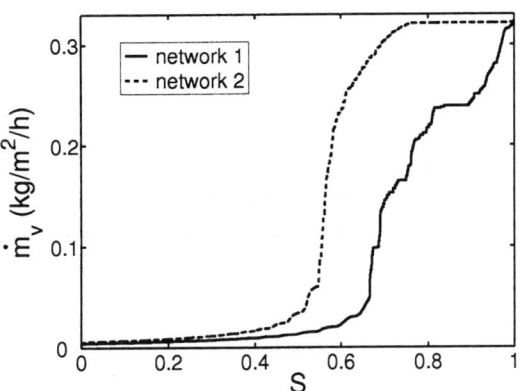

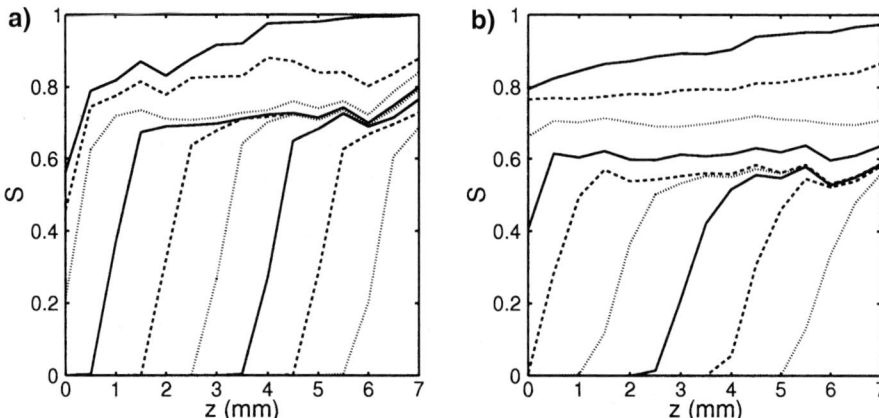

Fig. 7 Moisture profiles for (**a**) network 1 and (**b**) network 2 as in Fig. 5, drawn for multiples of 10% in network saturation

pore size distributions with highly correlated pore structures, as in network 2, where large throats are forming long channels, drying rates can be significantly higher, and a first drying period is observed. As can be seen in Fig. 7, moisture gradients for network 1 are steeper than for network 2.

We found that drying simulations in two and three dimensions may lead to contradictory conclusions: in the two-dimensional version of network 2, for example, capillary flow paths are easily cut off by emptying large throats. Therefore, three-dimensional modelling is imperative when structural influences are to be investigated.

It is clear that these first findings from network simulations are quite different from the results of the one-dimensional capillary model in which the liquid is viscous and forms a continuous phase. These contradictions will be overcome and a more comprehensive view become possible only by our current activity of including viscosity into the network model.

The next crucial step is to extend the model to heat transfer, since real drying can rarely be described by an isothermal model; previous work by Plourde and Prat (2003) and Huinink et al. (2002) will serve as basis. Due to non-uniform temperature and therefore non-uniform saturation vapour pressure, such a model must be able to describe vapour condensation and not only evaporation. Only then will the network model account for all transport effects, and it will make sense to compare its results with those of a continuous model.

Concerning this comparison, it is still an open question how much "drying information" has to be incorporated into the effective parameters of networks. Will they be computed from gradients and fluxes as observed during drying (cf. Nowicki et al. 1992), or from separate numerical experiments, i.e. by applying gradients to a network with a given saturation? And, in the latter case, according to which rules will the spatial distribution of liquid for this saturation be chosen: randomly or as observed during drying (cf. Segura and Toledo 2005)? Another important issue in this context is the network size needed to be representative, or the number of small network simulations needed to get average behaviour. In this frame, it might be reasonable to introduce periodic boundary conditions.

In order to investigate the influence of structure more comprehensively, we are currently looking into random pore networks, in which not only throat radius is chosen at random but also locations of pore nodes and existence of throat connections, leading to a distributed coordination number. Furthermore, spatial correlations have to be modelled. The aim will be to discover which structural parameters are sufficient to characterize the porous medium with respect to its drying behaviour; in this sense, equivalence classes of networks might be defined and algorithms found to generate networks representative of these classes.

3 Discrete element modelling: mixing and heat transfer in mechanically agitated beds

From single particles we now turn to granular material, i.e. large ensembles of particles. Again, our aim is to go beyond average behaviour of the whole system, but rather describe the properties of individual particles.

At present, when contact drying of particulate material in mechanically agitated beds is to be modelled, heat transfer from walls is usually described by a penetration model (Schlünder 1983; Schlünder and Mollekopf, 1984). In this traditional approach, the bed of particles is viewed as a continuum with effective properties, and the continuous mixing and heating of particles is discretised by alternating static periods of duration t_R with pure heat conduction to a stagnant bed and instantaneous perfect mixing thereafter. In a rotating drum, the time t_R is obtained from

$$t_R = t_{mix} \cdot N_{mix}, \quad t_{mix} = 1/n \tag{4}$$

where n is the rotational frequency and N_{mix} the number of revolutions needed for perfect mixing. The overall (time-averaged) heat transfer coefficient α is computed from

$$\frac{1}{\alpha} = \frac{1}{\alpha_{ws}} + \frac{1}{\alpha_{bed}} \tag{5}$$

where α_{ws} is the wall-to-bed transfer coefficient—as calculated from kinetic gas theory in the gap between wall and first layer of particles—and

$$\alpha_{bed} = \frac{2}{\sqrt{\pi}} \frac{\sqrt{\rho c_p \lambda_{bed}}}{\sqrt{t_R}} \tag{6}$$

is the penetration coefficient—as obtained from time-averaged heat conduction by Fourier's law with effective volumetric heat capacity ρc_p and effective thermal conductivity λ_{bed}.

Empirical correlations based on atmospheric (Schlünder 1980) and vacuum contact (Wunschmann 1974) drying experiments are available for the mixing number N_{mix}, relating it to rotational frequency and apparatus diameter. But it is clear that this single parameter of the penetration model might not be sufficient to grasp the complex motion of particles. Note that, for rotating drums, there are up to six different regimes of particle motion, depending on particle diameter, drum diameter, rotational frequency and loading. The rolling and the slumping regime were investigated in the presented work.

We will now discuss how the discrete element method, introduced by Cundall and Strack (1979) and implemented in PFC software by ITASCA, can be used as a modern alternative to test and upgrade penetration models—exemplified for the specific case of a horizontally rotating drum. In the discrete element approach the motion of individual particles is tracked by alternately computing forces between particles that are in contact and updating their velocities accordingly. Contact forces are computed from relative particle velocities, "overlap" of the particles and material properties describing stiffness, viscous damping and frictional slip.

At first we employed this purely mechanical model to study mixing of particles in a drum by two-dimensional simulations (Saage et al. 2005; Kwapinska et al. 2006). The authors have shown by comparison with experimental data from the literature (van Puyvelde et al. 1999, 2000) that the discrete element method has the potential for realistically describing the mixing behaviour (in respect to the character of mixing dynamics, to all trends and dependencies). A typical result is shown in Fig. 8, where spherical particles with a narrow size distribution are mixed; two (initially separated) fractions of particles, which differ only in their colour, are used to describe the extent of mixing. After a certain time, the number of contacts between particles of different colour has attained a constant level, apart from statistical fluctuations. This is the time needed for perfect mixing and can be directly compared with the parameter t_R of the penetration model. By discrete element modelling, mixing numbers N_{mix} were obtained, which are in average by a factor of 2.5 lower than those computed from the empirical correlations. A new correlation was proposed, which additionally accounts for drum loading by replacing the rotational frequency of the drum by that of the bed (for the rolling regime). Some examples of mixing times from penetration theory (t_R) and from discrete element modelling (t'_R) are given in Table 1. The mixing times t'_R are also converted into penetration coefficients α'_{bed} by Eq. (6).

Next, heat transfer was included into discrete element modelling (Kwapinska et al. 2005); to this end, a heat reservoir is associated to each particle and a thermal pipe to each contact between two particles or wall and particle (see Fig. 9). These thermal pipes are characterized by thermal resistances, obtained from heat conduction through the gas gap between the two contact partners (solid bridges and radiation are neglected); from geometric considerations, the wall-to-particle resistance is half of the particle-to-particle resistance and calculated as proposed by Schlünder (1983). The momentary heat flow rate is then computed from temperature differences. All simulations were performed for a constant mechanical and thermal time step of $2\,\mu s$. This allows us to fulfill the requirement that, within one time step, a change of particle temperature does not propagate further than a particle's immediate neighbours. A

Table 1 Comparison of simulation results with penetration theory, for 8 mm particles in a drum of diameter 0.25 m

		Penetration theory				Penetration theory with t'_R from DEM			DEM
Drum velocity, rpm	Drum load, %	t_R, s	α_{ws}, $\frac{W}{m^2 K}$	α_{bed}, $\frac{W}{m^2 K}$	α, $\frac{W}{m^2 K}$	t'_R, s	α'_{bed}, $\frac{W}{m^2 K}$	α', $\frac{W}{m^2 K}$	α_{sim}, $\frac{W}{m^2 K}$
28.6	15	21.7	70.5	181.0	50.7	3.10	478.9	61.4	68.2
287	30	5.46	70.5	360.8	58.9	3.16	474.3	61.3	69.2
28.6	30	21.7	70.5	181.0	50.7	7.30	312.1	57.5	67.7
4.78	30	61.8	70.5	107.2	42.5	19.25	192.2	51.6	66.3

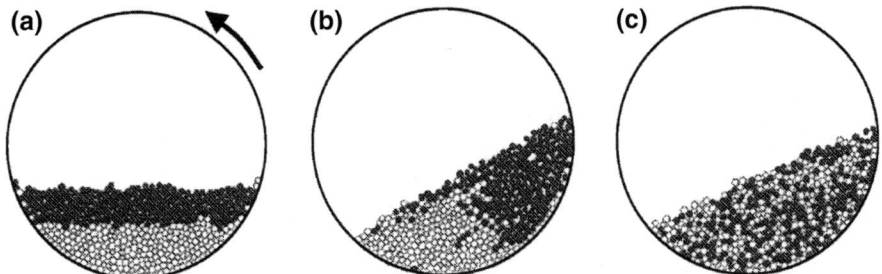

Fig. 8 Examples of different stages of mixing (in the rolling regime): (**a**) initial bed configuration, (**b**) bed after one drum revolution and (**c**) steady state

Fig. 9 Active thermal pipes for mono-sized spheres in a drum dryer

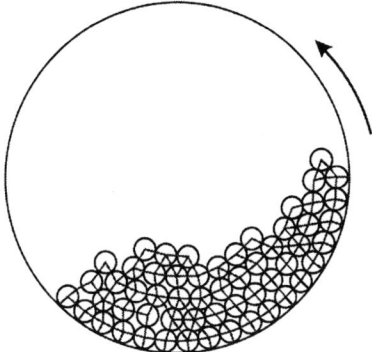

thermal pipe is made active if the respective two particles touch, i.e. if their overlap is greater than or equal to zero. After each time step, the activity status of each pipe is updated.

Computational time restricts the application of thermal discrete elements to a relatively small number of particles, so that the present investigation is of an exploratory character. Even for a maximum number of particles of 370, computational times of around 10 h on a commercial PC were necessary for about 70 s of simulated process time.

Our aim was to study heating of an agitated bed of mono-sized spheres in a rotating drum of constant wall temperature. However, we decided first to look at the limiting case when heat transfer is controlled by the wall-to-bed transfer coefficient ($\alpha \approx \alpha_{ws}$); temperature differences within the bed are then negligible and mixing does not play a role in overall heat transfer; to this end, the particle-to-particle thermal resistance was set to a value close to zero. The evolution of average bed temperature as obtained from discrete element modelling was found to coincide with the analytical solution.

Having confirmed that wall-to-bed heat transfer is correctly described, we then turned to heat transfer within the bed, which, in the penetration model, is characterised by the effective thermal conductivity of the bed λ_{bed}. If discrete element and penetration model are to be compared at full scale, this parameter must first be estimated from discrete element simulations (the remaining volumetric heat capacity is easily obtained). To this purpose, heat transfer from the drum to a stagnant bed was computed with the true resistances for the thermal pipes. In the absence of mixing, t_R in Eq. (6) is replaced by the real time and λ_{bed} is the only model parameter, which can be obtained by fitting the analytical solution to the discrete element result and has a value of $0.44\,\mathrm{W\,m^{-1}\,K^{-1}}$.

Finally, the combined mixing and heat transfer for a rotating drum were simulated and overall heat transfer coefficients α_{sim} estimated from the discrete element results. When comparing these to α' as predicted by the penetration model (with α_{ws} and λ_{bed} as by discrete elements), one can state that the penetration model underestimates the enhancement of heat transfer by mixing (see Table 1). Even the mixing times that had been obtained from purely mechanical simulations were too long to explain the difference, suggesting that it might not be feasible to describe the complex interaction of mixing and heat transfer by a simple penetration model.

A possible explanation for this is that the perfect mixing in the penetration model is assumed to be instantaneous and therefore "unstructured", whereas highly correlated, fluid-like particle motion can be observed in discrete element simulations (Kwapinska et al. 2005). However, another possible reason for the discrepancy will first have to be excluded: the current restriction to two dimensions and to only a few particles as compared to the experiments on which traditional correlations are based.

For illustration, we give some simulation results for combined mixing and heat transfer. Figure 10 shows the influence of mixing on the evolution of particle temperature: for a stagnant bed, the first layer of particles on the wall heats up rapidly, but the average bed temperature rises only slowly, whereas, for the rotating drum, temperature differences in the bed are leveled out. This is because — for the presented examples — contact resistance prevails over penetration resistance in the serial coupling of Eq. (5). The evolution of the particle temperature distribution is plotted in Fig. 11.

As the presented work shows, discrete element modelling has the potential for realistically describing simultaneous mixing and heat transfer in agitated beds. In the near future, the model will be extended to mass transfer so that all phenomena relevant in contact drying are accounted for. This will allow a detailed analysis and revision of existing drying models.

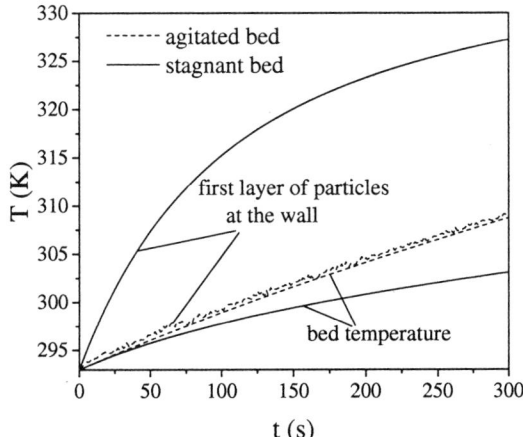

Fig. 10 Evolution of particle temperature in a drum heated by the wall (224 particles of diameter 8 mm, drum diameter 0.25 m, rotational frequency of drum 28.6 rpm, drum loading 30%)

Fig. 11 Evolution of particle temperature distribution in a rotation drum heated by the wall (particle diameter 8 mm, drum diameter 0.25 m, rotational frequency of drum 28.6 rpm)

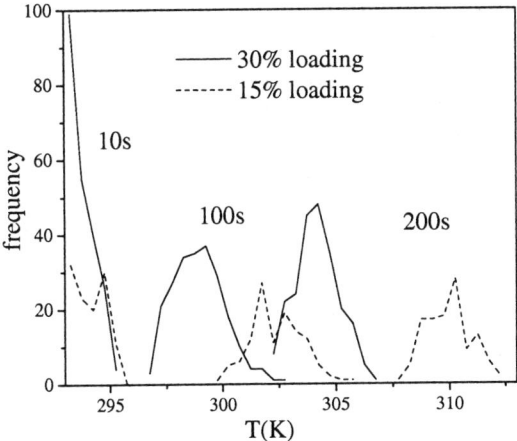

4 Population balances: distributed product quality in fluidised beds

Solids processing of particulate materials in fluidised beds, such as drying or spray granulation, are of major interest in the chemical industry. Various parameters—for material behaviour, apparatus design and operating conditions—influence the resulting products in terms of, for example, particle size, morphology and composition. Such disperse systems can best be modelled by population balances; these describe the temporal change of number density distributions of individuals of the disperse phase with respect to different external and internal coordinates.

Population balances have already been used in the past. However, these formulations were limited to relatively simple, one-dimensional models. The description of innovative and often very complex processes and operations for the production of granular materials as well as the quantification of important factors influencing product quality will require a multidimensional approach.

An example of applying population balances is continuous fluidised bed drying. The traditional approach to modelling this operation considers the disperse solids as one phase with average properties like particle size, moisture content and enthalpy. Recently, Burgschweiger (2000) proved that these models are generally suited for continuously operated dryers, but it was found, by comparison with experimental data, that such a model tends to underestimate the outlet moisture content (see Fig. 12).

Model accuracy can be significantly improved by use of population dynamics. Due to different retention times of particles, a moisture distribution in the product stream is expected. This can be modelled by introducing an additional property coordinate, the age τ of particles. The temporal evolution of the number density n for an ideally mixed system is then given by

$$\frac{\partial n(t,\tau)}{\partial t} = -\frac{\partial n(t,\tau)}{\partial \tau} + \dot{n}_{\text{in}}(t) - \dot{n}_{\text{out}}(t,\tau) \tag{7}$$

where $\dot{n}_{\text{in}}(t)$ indicates the inlet rate of particles and $\dot{n}_{\text{out}}(t,\tau)$ the particle flow rate at the outlet. The expression $\partial n(t,\tau)/\partial \tau$ represents the aging of particles. Integration of Eq. (7) results in the age distribution function of particles. Since we are not only interested in this particular aging process, but in the moisture content distribution,

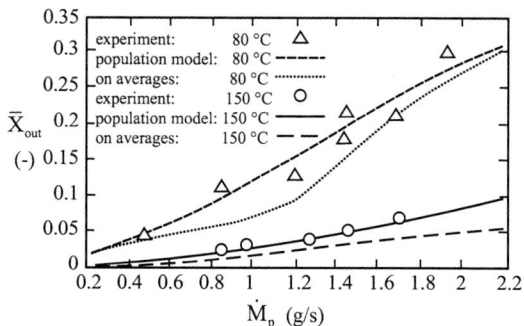

Fig. 12 Static continuous fluidised bed drying: measured and simulated (on averages and population balance) outlet moisture content of γ-Al_2O_3 particles (ϕ 1.8 mm, initial moisture content 0.60–0.65) for different particle flow rates

which is a significant parameter of product quality, we also need to apply the concept of population balances to the amount of liquid contained in the particles and to particle enthalpy. The evolution of the above number distribution will not be influenced by interaction with the gas phase. But as soon as processes such as drying and heating of particles are involved, additional sinks and sources have to be introduced in the liquid mass and enthalpy balances of the particles, respectively. A model extended in this way yields considerably better values for product moisture at the dryer outlet, as can be seen in Fig. 12 (Burgschweiger and Tsotsas 2002).

Extensive experiments are currently being carried out, in which the distribution of outlet moisture is determined by gravimetric measurement of single particles. In Fig. 13, the measured normalized particle density distribution q_0 is shown as a function of single particle moisture content. This result verifies that the outlet moisture of mono-sized particles is distributed over a relatively wide range, reflecting different retention times in the apparatus. As a consequence of this, it seems to be reasonable to find new modelling approaches, such as population balances, that have the potential to replace the existing models, which are based on averages.

Following this approach, the set of product characteristics can be arbitrarily extended to other (outlet) properties, such as colour or bioactivity, which depend on the history of the particle related to temperature and water activity.

A second research field is related to processes of particle formulation, such as particle coating and spray granulation (agglomeration), whereas the latter is the scientifically more interesting and more complex case. For these processes,

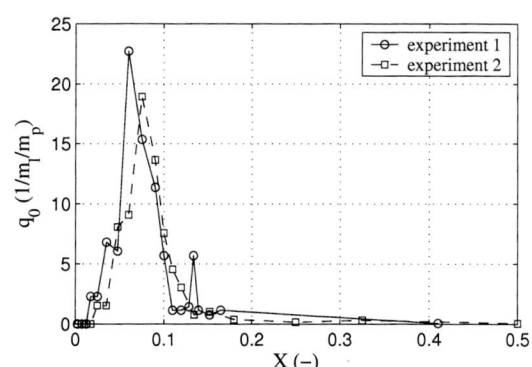

Fig. 13 Static continuous fluidised bed drying: experimental moisture content of γ-Al_2O_3 particles (ϕ 1.8 mm, initial moisture content 0.60–0.65) at dryer outlet

one-dimensional population balances have been applied in numerous approaches, whereby the particle volume v has been considered as the only significant property.

In fact, other relevant coordinates of the particle distribution are amount of liquid l and enthalpy. Assuming binary agglomeration events, the temporal change of the number density f can be described as

$$\frac{\partial f(t,v,l)}{\partial t} + \frac{\partial G(v,l) \cdot f(t,v,l)}{\partial l}$$
$$= \frac{1}{2} \int_0^v \int_0^l \beta(t, v-\varepsilon, \varepsilon, l-\gamma, \gamma) \cdot f(t, v-\varepsilon, l-\gamma) \cdot f(t, \varepsilon, \gamma) \mathrm{d}\gamma \mathrm{d}\varepsilon$$
$$- \int_0^\infty \int_0^\infty \beta(t, v, \varepsilon, l, \gamma) \cdot f(t, v, l) \cdot f(t, \varepsilon, \gamma) \mathrm{d}\gamma \mathrm{d}\varepsilon \qquad (8)$$

where G is the combined rate of wetting and drying—both depending on particle size, the second additionally on moisture content—and β describes the kinetics of the agglomeration process in terms of a kernel. The left term of this equation describes the accumulation of particles of certain size and amount of liquid. The first integral on the right side depicts the birth and the second integral represents the death of particles due to agglomeration. Since we have two properties of the solid phase (size and amount of liquid), double integration becomes necessary. Here ε and γ are variables of integration for the coordinates particle size and liquid mass, respectively. For simplicity, the coordinate of enthalpy is not elaborated here. Otherwise three integrals in the population balance would have been needed. Furthermore, one would have to introduce one additional advection term (such as for the drying and wetting) that takes into account the heat and enthalpy fluxes between solid and gas phase.

The solution of such multidimensional population balances requires, on the one hand, a great numerical effort (exponential in the number of particle properties) and, on the other hand, a theoretical expression for the multidimensional kernel. The marginal distributions approach offers a way to reduce the model, if the agglomeration process is assumed to depend only on particle size and not on moisture content, namely $\beta(t, v, \varepsilon, l, \gamma) = \beta(t, v, \varepsilon)$. By integrating over the amount of liquid l, we first define number density n and the liquid mass density m_l as

$$n(t, v) = \int_0^\infty f(t, v, l) \mathrm{d}l \qquad (9)$$

$$m_l(t, v) = \int_0^\infty l \cdot f(t, v, l) \mathrm{d}l \qquad (10)$$

Then, the multidimensional population balance of Eq. (8) may be rewritten as several one-dimensional balances, one for the number distribution

$$\frac{\partial n(v)}{\partial t} = \frac{1}{2} \int_0^v \beta(u, v-u) \cdot n(u) \cdot n(v-u) \mathrm{d}u$$
$$- n(v) \int_0^\infty \beta(u, v) \cdot n(u) \mathrm{d}u \qquad (11)$$

and one for the moisture distribution

$$\frac{\partial m_l(v)}{\partial t} = \int_0^v \beta(u, v-u) \cdot n(u) \cdot m_l(v-u) \mathrm{d}u$$
$$- n(v) \int_0^\infty \beta(u, v) \cdot m_l(u) \mathrm{d}u + \frac{\partial}{\partial v} \left(\dot{M}_{\mathrm{np}} - \dot{M}_{\mathrm{ps}} \right) \qquad (12)$$

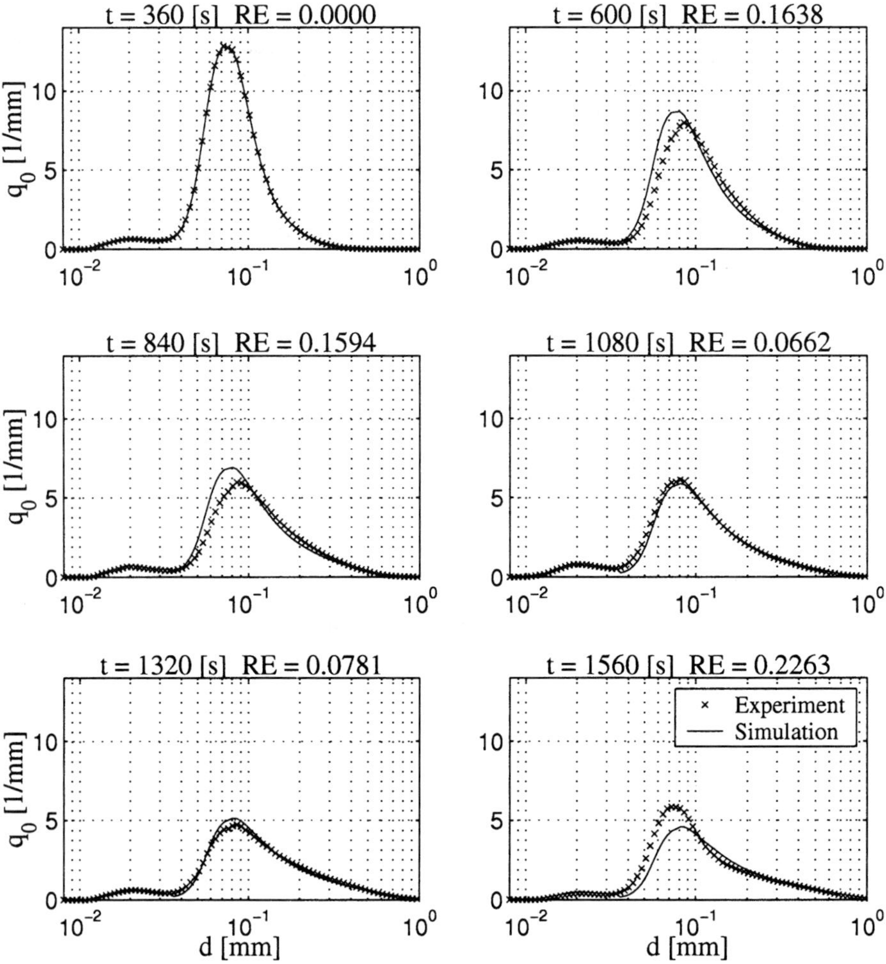

Fig. 14 Evolution of measured and simulated particle size distribution in fluidised bed spray agglomeration (RE = cumulative relative error)

where $\dot{M}_{np}$ is liquid mass flow from the spray nozzle and $\dot{M}_{ps}$ drying rate of the particles of size v. These terms can be deduced directly from the advection term on the left side of Eq. (8). As a result of this one obtains wetting and drying rates that depend on particle size but not on moisture content. (Again, the balance for the enthalpy distribution is not given here.) By this model reduction, the numerical effort is considerably reduced (linear in the number of properties).

The process modelled by this equation has been investigated by agglomeration experiments on microcrystalline cellulose in a fluidised bed (Peglow 2005). The governing agglomeration kernel was estimated from experimental data taking into account size-dependent nuclei formation. For this purpose, a new inverse technique was derived (Peglow et al. 2006) based on work of Bramley et al. (1996). An example for the good correspondence of model and measurement data is given in Fig. 14.

Based on work of Hounslow et al. (2001), new discretisation methods of the continuous population balance equation were developed (Peglow et al. 2006a) to solve this system. These new discretisation algorithms allow the prediction of intensive properties (temperature and moisture content) in addition to extensive properties of the disperse phase, such as size, enthalpy and amount of water.

As indicated in Eq. (12), the agglomeration process in the disperse phase was not considered in an isolated manner, but in the frame of a heterogeneous fluidised bed model with heat and mass transfer between continuous and disperse phase. More specifically, the drying rate was treated as dependent on particle size as well as on moisture content and temperature associated with this particle size.

In all existing models, the agglomeration kernel is a function of particle size only. However, other particle properties have a strong influence on the agglomeration process so that successful and first principle modelling needs to be multidimensional. As most important properties, one may consider size and moisture content of granules, but also porosity and binder content.

In a first step, the amount of liquid l in a single particle will be introduced as a full coordinate as shown in Eq. (8). Then, the agglomeration kernel β as well as the drying rate G will depend on this property, the latter establishing a direct link of mass transfer processes between granular and gas phase.

To get a better understanding of the structure of the kernel, experimental data for the temporal change of the two-dimensional distribution (size and water content) are needed. Such measurements are very demanding, but first trials with magnetic suspension gravimetry and NMR methods are promising.

5 Conclusion

Recent approaches to drying modelling have been presented. Unlike traditional techniques, relevant quantities are modelled and measured locally and not in terms of averages. In this way, processes at the pore scale (for single particles) and for individual particles (for beds of particles) are not masked. Therefore, new insight in the drying processes at different scales may well be expected.

Acknowledgements The authors would like to state that part of the presented work has been financed by the Graduiertenkolleg 828 "Micro-Macro-Interactions in Structured Media and Particle Systems"(funded by the German Research Foundation) and by the State of Saxony-Anhalt.

References

Bramley, A.S., Hounslow, M.J., Ryall, R.L.: Aggregation during precipitation from solution: A method for extracting rates from experimental data. J. Colloid Interface Sci. **183**, 155–165 (1996)

Burgschweiger, J.: Modellierung des statischen und dynamischen Verhaltens von kontinuierlich betriebenen Wirbelschichttrocknern. Dissertation, Universität Magdeburg, VDI-Verlag, Reihe 3, Nr. 665 (2000)

Burgschweiger, J., Tsotsas, E.: Experimental investigation and modelling of continuous fluidized bed drying under steady-state and dynamic conditions. Chem. Eng. Sci. **57**, 5021–5038 (2002)

Cundall, P.A., Strack, O.D.L.: A discrete numerical model for granular assemblies. Geotechnique **29**, 47–65 (1979)

Hounslow, M.J., Pearson, J.M.K., Instone, T.: Tracer studies of high shear granulation: II. Population balance modeling. AIChE J. **47**, 1984–1999 (2001)

Huinink, H.P., Pel, L., Michels, M.A.J., Prat, M.: Drying processes in the presence of temperature gradients — pore-scale modelling. Eur. Phys. J. E. **9**, 487–498 (2002)

Irawan, A., Metzger, T., Tsotsas, E.: Pore network modelling of drying: combination with a boundary layer model to capture the first drying period. Proc. 7th WCCE, Glasgow, United Kingdom (2005)

Kwapinska, M., Saage, G., Tsotsas, E.: On the way from penetration models to discrete element simulations of contact dryers. XI Polish Drying Symposium, Poznan, Poland (2005)

Kwapinska, M., Saage, G., Tsotsas, E.: Mixing of particles in rotary drums: A comparison of discrete element simulations with experimental results and penetration models for thermal processes. Powder Technol. **161**, 69–78 (2006)

Laurindo, J.B., Prat, M.: Numerical and experimental network study of evaporation in capillary porous media. Chem. Eng. Sci. **53**, 2257–2269 (1998)

Metzger, T., Tsotsas, E.: Influence of pore size distribution on drying kinetics: a simple capillary model. Drying Technol. **23**, 1797–1809 (2005)

Metzger, T., Irawan, A., Tsotsas, E.: Discrete modeling of drying kinetics of porous media. Proc. 3rd Nordic Drying Conference, Karlstad, Sweden (2005)

Nowicki, S.C., Davis, H.T., Scriven, L.E.: Microscopic determination of transport parameters in drying porous media. Dry. Technol. **10**, 925–946 (1992)

Peglow, M.: Beitrag zur Modellbildung eigenschaftsverteilter Feststoffsysteme am Beispiel der Wirbelschicht-Sprühagglomeration. Dissertation, Universität Magdeburg (2005)

Peglow, M., Kumar, J., Warnecke, G., Heinrich, S., Mörl, L., Hounslow, M.J.: Improved discretized tracer mass distribution of Hounslow et al., AIChE J. **52**, 1326–1332 (2006a)

Peglow, M., Kumar, J., Warnecke, G., Heinrich, S., Mörl, L.: A new technique to determine rate constants for growth and agglomeration with size- and time-dependent nuclei formation. Chem. Eng. Sci. **61**, 282–292 (2006a)

Perré, P., Turner, I.W.: A 3-D version of TransPore: a comprehensive heat and mass transfer computational model for simulation the drying of porous media. Int. J. Heat Mass Transfer **42**, 4501–4521 (1999)

Plourde, F., Prat, M.: Pore network simulations of drying of capillary porous media. Influence of thermal gradients, Int. J. Heat Mass Transfer **46**, 1293–1307 (2003)

Saage, G., Kwapinska, M., Tsotsas, E.: Discrete element simulation of the mixing time of granular solids. In: Proc. 16th International Symposium on Trends in Applications of Mathematics to Mechanics (pp. 441–450). Shaker Verlag (2005)

Schlünder, E.U.: Der Wärmeübergang an ruhende, bewegte und durchwirbelte Schüttschichten. VT-Verfahrenstechnik **14**, 459–468 (1980)

Schlünder, E.U.: Heat transfer to packed and stirred beds from the surface of immersed bodies. Chem. Eng. Process. **18**, 31–53 (1983)

Schlünder, E.U., Mollekopf, N.: Vacuum contact drying of free flowing, mechanically agitated particulate material. Chem. Eng. Process. **18**, 93–111 (1984)

Segura, L.A., Toledo, P.G.: Pore-level modeling of isothermal drying of pore networks. Effects of gravity and pore shape and size distributions. Chem. Eng. J. **111**, 237–252 (2005)

van Puyvelde, D.R., Young, B.R., Wilson, M.A., Schmidt, S.J.: Experimental determination of transverse mixing kinetics in a rolling drum by image analysis. Powder Technol. **106**, 183–191 (1999)

van Puyvelde, D.R., Young, B.R., Wilson, M.A., Schmidt, S.J.: Modelling transverse mixing kinetics in a rolling drum. CJChE **78**, 635–642 (2000)

Wunschmann, J.: Wärmeübertragung von beheizten Flächen an bewegte Schüttungen bei Normaldruck und Vakuum. Dissertation, Universität Karlsruhe (1974)

Yiotis, A.G., Stubos, A.K., Boudouvis, A.G., Yortsos, Y.C.: A 2-D pore network model of the drying of single-component liquids in porous media. Adv. Water Resour. **24**, 439–460 (2001)

Yiotis, A.G., Stubos, A.K., Boudouvis, A.G., Tsimpanogiannis, I.N., Yortsos, Y.C.: Pore-network modeling of isothermal drying in porous media. Trans. Porous Media **58**, 63–86 (2005)

Non-linear heat and mass transfer during convective drying of kaolin cylinder under non-steady conditions

Musielak Grzegorz · Banaszak Jacek

Received: 15 November 2005 / Accepted: 26 March 2006 /
Published online: 30 August 2006
© Springer Science+Business Media B.V. 2006

Abstract The self-consistent model of heat and mass transfer during convective drying of capillary porous media describing both the first and the second periods of drying is presented in Musielak (Wydawnictwo Politechniki Poznańskiej, seria Rozprawy, nr 386 (2004a); Chem. Process Eng. **25**, 393–409 (2004b)). The results of simulations of processes in steady conditions are shown (Musielak Wydawnictwo Politechniki Poznańskiej, seria Rozprawy, nr 386 (2004a); Chem. Process Eng. **25**, 393–409 (2004b)). The main aim of the present work is to compare experimental results with those from numerical simulations. Three convective drying processes have been performed experimentally. The first and the second periods of drying are considered, during which the humidity of air in the dryer changes due to evaporation. The first process is used to establish drying parameters, whereafter the two remaining processes are simulated. Good agreement between experimental and simulation results is found, both qualitative and quantitative.

Keywords Kaolin clay · Non-steady drying conditions · Convective drying · Modeling · Experiment · Simulation

1 Introduction

According to Pabis (2005), "mathematical models of processes of drying solid bodies can [...] be either empirical formulas or theoretical models". The empirical formulas result from the existence of some empirical regularity of the investigated systems. They describe the regularity. Usually, they are constructed with the help of dimensional analysis and the theory of similarity. A new way of obtaining the empirical

---

M. Grzegorz (✉) · B. Jacek
Insitute of Technology and Chemical Engineering, Poznań University of Technology, Pl. Marii Skłodowskiej-Curie 2, 60-965 Poznań, Poland
e-mail: Grzegorz.Musielak@put.poznan.pl

B. Jacek
e-mail: Jacek.Banaszak@put.poznan.pl

formulas is connected with the neural network method. Empirical models have some limited importance for technology. They are also useful for testing theoretical models. The empirical formulas are also used as part of a mixed mathematical model. Generally, however, the empirical models give no scientific explanation of any physical phenomenon.

Theoretical models of drying can be divided into two groups: global models and structural ones. All global models treat the dried body as an open thermodynamic system and the drying medium as a surrounding of the system. The main advantage of global models is their simplicity. They need a knowledge of few parameters. It is then easy to calculate both the mass and temperature of the dried body. Global models are therefore widely used in technology. But global models also have some disadvantages. For instance, one question is, what is the average temperature? Another problem is connected with the critical moisture content, X_{kr}. In global models, this is a function of dried material, shape and size of dried body, and drying conditions. This must therefore be specified experimentally for each drying process.

The aforementioned and others disadvantages can be elucidated with the help of structural models, which treat the dried body as a continuum. The most commonly used models are due to Philip and De Vries (1957), Luikov (1975), Whitaker (1977), and Kowalski (2003). All of them are non-linear. All could be simplified to the Fick-type equation of diffusion (linear or non-linear). In all of them, the problem is to establish their coefficients. As Coumans (2000) said: "(Models) require quite a number of model parameters to be found experimentally [...] (therefore) are not easy to use in practice". Another disadvantage of structural models is that they lead to sets of coupled non-linear partial differential equations. The structural model used in this work is based on the thermomechanical model of heat and mass transfer during convective drying of capillary porous media (Kowalski 1996, 2003). The mathematical model has been verified and adapted for both the first and the second periods of drying (Musielak 2004a, b). The model is self-consistent.

The main aim of the present work is the validation of the model, to which end, a comparison of experimental results with numerical simulations for convective drying of a kaolin clay sample was done. The experiments were carried out at three different temperatures in a laboratory drier: about 40, 50, and 60°C. The first drying process 40°C, was performed to establish drying parameters. The results of the simulations of the remaining two processes are presented to show the ability of the model used to reflect the real drying processes and to validate the model.

2 Mathematical model

The present work is devoted to heat and mass transfer during drying of a kaolin cylinder (see Fig. 1)

The mathematical model of the transfer phenomena (Musielak 2004a, b) consists of the moisture contents and energy balance equations and of relations describing fluxes of moisture and heat inside dried material and fluxes of the moisture and heat outside the material due to convective drying.

The mass balance equation is in the form

$$\frac{\partial X}{\partial t} = -\frac{1}{\rho^{(s)}} \mathrm{div}\, \mathbf{j}^{(m)}, \qquad (1)$$

Fig. 1 Dried cylinder

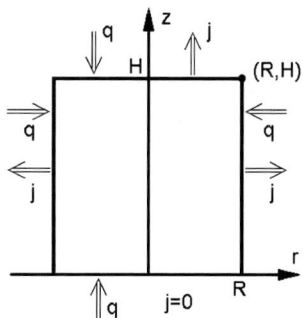

where X is the mass of moisture related to the mass of the skeleton, $\rho^{(s)}$ is the partial density of the skeleton and $\mathbf{j}^{(m)}$ is the moisture flux. The flux is assumed to be proportional to the gradient of moisture content potential (Kowalski 1996, 2003). Due to small deformations of dried kaolin the flux is a function of the moisture content X and the temperature T

$$\mathbf{j}^{(m)} = -\Lambda_X \left(\frac{c^{(X)}}{\rho^{(s)}} \operatorname{grad} X + \frac{c^{(T)}}{\rho^{(s)}} \operatorname{grad} T \right), \qquad (2)$$

where Λ_X is the moisture transfer coefficient, $c^{(X)}$ and $c^{(T)}$ are the moisture contents and the temperature coefficients of the moisture potential, respectively. The assumption that the moisture transport participates mainly in the form of liquid and the transport of vapor inside of the material can be omitted gives the moisture transfer coefficient Λ_X in the form (Musielak 2000)

$$\Lambda_X = \frac{K}{\eta} \left(\rho_R^{(l)} \right)^2, \qquad (3)$$

where K is the effective geometric permeability, η is the dynamic viscosity and $\rho_r^{(l)}$ is the real mass density of the liquid phase inside the pores. The effective permeability K is a function of the moisture content. On the basis of the work of Ketelaars (1992) and Augier et al. (2000, 2002) the effective permeability of kaolin is taken in the form

$$K = \begin{cases} K_0 & \text{for} \quad X > X_{kr}, \\ A * 10^{B*X-C} & \text{for} \quad X \le X_{kr}, \end{cases} \qquad (4)$$

where X_{kr} denotes the critical moisture content (locally defined, see Musielak 2004a, b).

The boundary conditions for the convectively dried cylinder (Fig. 1) result directly from the mass balance on the boundary

$$\mathbf{j}^{(m)}\big|_{\partial\Omega^-} \cdot \mathbf{n} = j\big|_{\partial\Omega^+}, \qquad (5)$$

where the moisture flux inside the material $\mathbf{j}^{(m)}$ is equal to the flux of the evaporated mass j. The $\partial\Omega^-$ and $\partial\Omega^+$ denote the boundary of the material from the inside and outside, respectively, and $\mathbf{n}$ is the outside normal to the boundary. There is no mass evaporation on the base of the cylinder (Fig. 1), therefore

$$\mathbf{j}^{(m)}\big|_{z=0} = 0. \qquad (6)$$

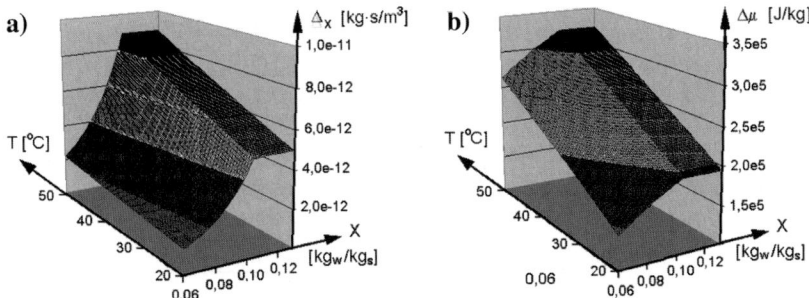

Fig. 2 Dependence of the moisture transfer coefficient Λ_X and the difference of the potentials $\Delta\mu$ on temperature T and on moisture content $X|_{\partial\Omega^-}$ of the material surface

The flux of evaporated mass j is assumed to be proportional to the difference between the chemical potential of humid air near the surface of the dried body $\mu|_{\partial\Omega^+}$ and the chemical potential of ambient air μ_a (Kowalski and Strumiłło 1997)

$$j|_{\partial\Omega^+} = \alpha^{(m)} \left(\mu|_{\partial\Omega^+} - \mu_a \right) = \alpha^{(m)} \Delta\mu, \qquad (7)$$

where $\alpha^{(m)}$ is the coefficient of convective mass transfer between dried material and surroundings. The difference of the potentials is in the form (Musielak 2004a, b)

$$\mu|_{\partial\Omega^+} - \mu_a = s \left(T_a - T|_{\partial\Omega^+} \right) + R \left(T|_{\partial\Omega^+} \ln \left(\varphi|_{\partial\Omega^+} \frac{p_s^{(v)}|_{\partial\Omega^+}}{p} \right) \right.$$
$$\left. - T_a \ln \left(\varphi_a \frac{p_{sa}^{(v)}}{p} \right) \right), \qquad (8)$$

where s is the vapour entropy per unit mass, T is the absolute temperature, p is the total pressure, $p_s^{(v)}$ is the partial pressure of saturated vapor and φ is the relative humidity of air. It is assumed that the relative humidity near the surface of the dried material $\varphi|_{\partial\Omega^+}$ is a function of the moisture content on the surface of the material $X|_{\partial\Omega^-}$. The linear dependence of the function is assumed

$$\varphi|_{\partial\Omega^+} = \begin{cases} 1 & \text{for} \quad X|_{\partial\Omega^-} > X_{kr} \\ 1 - \frac{1-\varphi_a}{X_{kr}-X_r} \left(X_{kr} X|_{\partial\Omega^-} \right) & \text{for} \quad X_{kr} \geq X|_{\partial\Omega^-} > X_r \\ \frac{\varphi_a}{X_r} X|_{\partial\Omega^-} & \text{for} \quad X|_{\partial\Omega^-} \leq X_r, \end{cases} \qquad (9)$$

where X_{kr} is the critical moisture contents and X_r is the equilibrium moisture contents. The relation (9) describes convective mass evaporation both during the first and the second periods of drying.

The Eqs. 2–9 are non-linear boundary condition for the mass transfer. The dependence of the moisture transfer coefficient Λ_X and the difference of the potentials $\Delta\mu$ on the temperature T and on the moisture content $X|_{\partial\Omega^-}$ of the material surface are shown in Fig. 2. The assumed parameters of the ambient air in that case are: $T_a = 50°C$, $\varphi_a = 0.115\,1$. For moisture contents X greater than the critical moisture contents X_{kr} both the moisture transfer coefficient Λ_X and the difference of the potentials $\Delta\mu$ are functions of temperature only, but for smaller moisture contents they are functions of temperature and moisture contents.

The energy balance equation is in the form

$$\rho^{(s)}\left(c_v^{(s)} + c_v^{(l)}X\right)\frac{\partial T}{\partial t} = -\operatorname{div}\mathbf{q}, \qquad (10)$$

where $c_v^{(s)}$, $c_v^{(l)}$ denote the specific heat of skeleton material and liquid, respectively, and the heat flux is expressed in the form (Musielak 2004b)

$$\mathbf{q} = -\left((1-f)\Lambda_\vartheta^{(s)} + f\frac{\rho^{(l)}}{\rho_0^{(l)}}\Lambda_\vartheta^{(l)}\right)\operatorname{grad} T, \qquad (11)$$

where $\Lambda_\vartheta^{(s)}$, $\Lambda_\vartheta^{(l)}$ are the thermal conductivies of skeleton material and liquid, respectively, f is the volume porosity, and the ratio of the liquid densities $\rho^{(l)}/\rho_0^{(l)}$ describes saturation of dried material. The energy balance on the boundary gives

$$\mathbf{q}|_{\partial\Omega^-}\cdot\mathbf{n} = q|_{\partial\Omega^+} + lj|_{\partial\Omega^+}, \qquad (12)$$

where q denotes the heat flux from the surroundings to the boundary and the second term on the right-hand side denotes the energy of evaporation on the boundary (l is the latent heat of evaporation). The convective heat flux q is taken in Newtonian form

$$q|_{\partial\Omega^+} = \alpha^{(\vartheta)}\left(T|_{\partial\Omega^+} - T_a\right), \qquad (13)$$

where $\alpha^{(\vartheta)}$ is the coefficient of convective heat transfer between dried material and surroundings.

The set of Eqs. 1–13 provides a self-consistent mathematical model of convective drying of capillary-porous media.

3 Experimental data

The studies were carried out on kaolin clay (KOC) made by the Surmin–Kaolin S.A. company. The chemical composition, structure and some physical properties of the KOC kaolin can be found on the producer's website (see the last reference). Kaolin is widely used in the ceramics industry, in manufacturing wall and floor tiles, tableware and sanitary ware. It provides the strength and plasticity in the shaping of these products and reduces the amount of pyroplastic deformation in the process of firing.

To prepare KOC for experiments some water was added to achieve an initial moisture content of approximately $X_0 = 0.45$. Kaolin was then homogenized and stored in closed box for 48 h to allow the moisture distribution in the whole material to become uniform. The mass obtained was used to mold cylindrically shaped samples (radius $R = 0.03$ m and height $H = 0.06$ m).

All experiments were carried out in the laboratory chamber dryer (Softmed SL 25). During the experiments, temperature T_a and relative humidity φ_a inside the chamber were measured every minute with the help of a Pt 100 probe and humidity sensor (DO 9,861 T Delta OHM), respectively, and transmitted into computer memory. The experiments were done in two steps. During the first one (Fig. 3a), three samples were dried to obtain drying curves. The kaolin samples were put on the aluminum table suspended from the electronic balance (Radwag 720), which registered the moisture removal in one-minute time periods with accuracy 0.01 g.

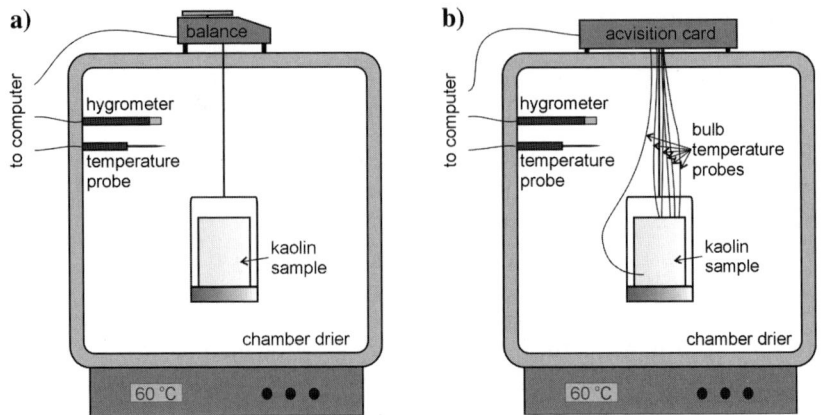

Fig. 3 Experimental set-up: (**a**) determination of drying kinetic curve of the sample, (**b**) determination of temperature inside the sample

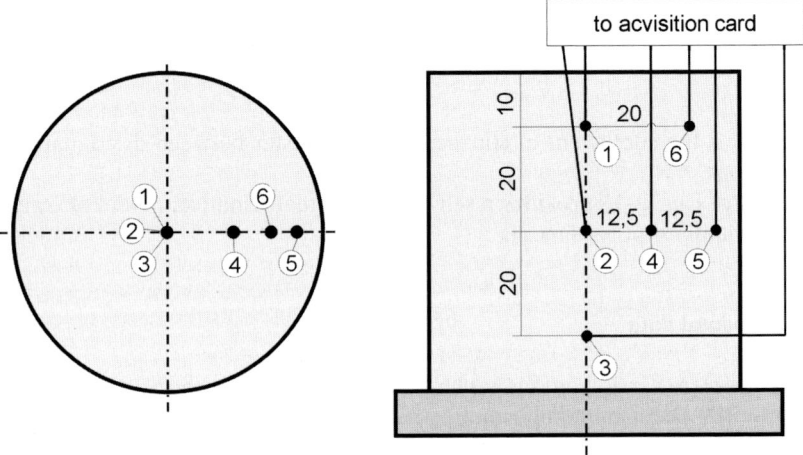

Fig. 4 Distribution of bulb temperature probes in kaolin sample

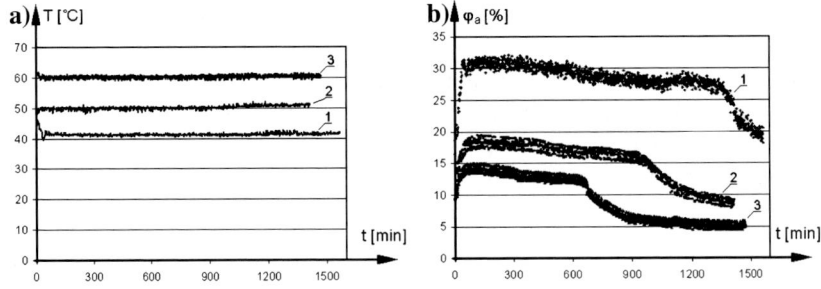

Fig. 5 Drying conditions during the experiments: (**a**) temperature in the chamber, (**b**) relative humidity in the chamber

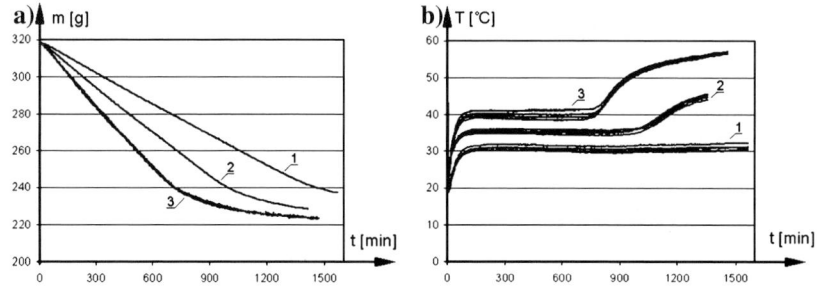

Fig. 6 Change of: (**a**) mass of the sample, (**b**) temperature inside the sample during the experiments

During the second step of the experiment another three samples were dried but this time the temperature inside the material was measured (Fig. 3b). The kaolin samples were placed in the laboratory dryer in exactly the same way as in the previous experiment. Ni–Cu bubble temperature probes of diameter 1 mm and accuracy 2% were used. Figure 4 shows the distribution of six probes placed in the sample. Three of them were put in the cylinder's axis of symmetry (1, 2, 3), another two along the radius at the half height of the kaolin sample (4, 5), and one near the upper surface (6). Another sensor was used to measure temperature in the chamber. The signals from all sensors were recorded at one-minute time intervals by the computer using software compatible with the canvassing card GPIB.

The experiments were performed at three temperatures of the drying medium: 41.5, 50.8, and 60.4°C (see Fig. 5a). As mentioned before, each experiment was done in two steps. Measurements of drying conditions in the chamber showed that the differences between them during these two steps were of little imprtance, so it is assumed that the conditions of drying for each respective temperature were the same in both steps of experiment.

All experiments took more than 24 h. The humidity of the drying medium changed during the experiments (see Fig. 5b). The main reason for the increase of humidity during the initial stage of drying was the evaporation of water from the sample. After that the humidity decreased slowly during the first period of drying and decreased rapidly during the second period. The slow exchange of the air in the drying chamber with air outside and the small volume of the chamber ($0.025\,\mathrm{m}^3$) were the reasons for the non-steady drying conditions.

The results of the experiments are presented in Fig. 6, which shows that there are three ways of observing of the passage between the first and the second periods of drying. The drying ratio is constant during the first period, so the change of the sample mass is a part of a straight line (see Fig. 6a). The start of the curvilinear part of the graph (Fig. 6a) is equivalent to the beginning of the second period of drying. It is impossible, though, to exactly fix the change from the first to the second period of drying from the drying curve (Fig. 6a).

The decrease of the drying ratio is due to decrease of the mass flux j (Eq. 5). This causes two phenomena. First, the decrease of the humidity inside the drying chamber (see Fig. 5b) and second, an increase of total flux $q|_{\partial\Omega^-}$ (Eq. 11) at the boundary, and as a result the increase of temperature inside the sample (see Fig. 6b). The second period of drying for the kaolin sample dried at 41.5°C began at about 1400 min, shortly before end of the experiment. Harsher drying conditions shortened the sample drying

Table 1 Constants describing skeleton and constants describing diffusion of moisture inside the material

Partial density of the skeleton	$\rho^{(s)}$	kg/m^3	1,248
Volume porosity	f	1	0.5
Thermal conductivity of the skeleton material	$\Lambda_\vartheta^{(s)}$	W/(m^{2}K)	1.78
Specific heat of the skeleton material	$c_v^{(s)}$	J/(kg K)	728.5
Moisture contents coefficient of the moisture potential	$c^{(X)}$	J/m^3	4.65×10^9
Temperature coefficient of the moisture potential	$c^{(T)}$	J/(m^{3}K)	2.64×10^6
Critical moisture contents	X_{kr}	kg$_w$/kg$_s$	0.1
Equilibrium moisture contents	X_r	kg$_w$/kg$_s$	0.005
Constant describing the effective permeability of kaolin (Eq. 4)	K_0	m^2	2.5×10^{-20}
Constant describing the effective permeability of kaolin (Eq. 4)	A	m^2	46.1
Constant describing the effective permeability of kaolin (Eq. 4)	B	kg$_s$/kg$_w$	14.10437
Constant describing the effective permeability of kaolin (Eq. 4)	C	1	22.6756

time. For drying at about 50 and 60°C the first period of drying changed into the second one at about 1,100 and 700 min, respectively.

During the first period of drying under steady state drying conditions the temperature inside the sample should be steady and equal to the wet bulb temperature. The theoretical values of wet bulb temperatures are 26.2, 27.5, and 31.4°C for three drying conditions considered (40.5, 50.8, and 60.4°C). In our experiments, we noticed higher values of temperature inside the samples: about 30, 35, and 40°C. The main reason for this is that moisture evaporation from the sample was rendered difficult due to the small volume of the laboratory chamber and small exchange of the drying medium inside the chamber with outside air. Such specific drying conditions generate a lower energy flux evaporated with steam from the sample surface than the theoretical value calculated for drying in free air, which leads to higher sample temperature than the theoretical wet bulb temperature, as expected and measured.

4 Numerical simulations

The main aim of the work is to compare experimental data and simulations for non-steady drying conditions. The changes in the conditions result mainly from passage between the first and the second periods of drying (see Fig. 5b). In the case of drying at temperature 41.5°C, the experiment comprised only the first period of drying and the start of the second one. The results of the experiment were used to estimate the coefficients of the convective mass and heat exchange. Therefore, only the simulations of drying at 50.8 and 60.4°C were performed.

The set of material coefficients used in calculation could be divided into several groups. The constants describing the skeleton and the constants describing diffusion of moisture inside the material are presented in Table 1. These constants are stated on the basis of literature reports and our own research (Comini and Lewis 1976; Strumiłło 1983; Ketelaars 1992; Augier et al 2000, 2002; Banaszak 2005; Musielak et al 2005).

The coefficients describing water in the material, water vapour and moist air are temperature dependent. All of them were tabulated in the range 0-100°C with a 1

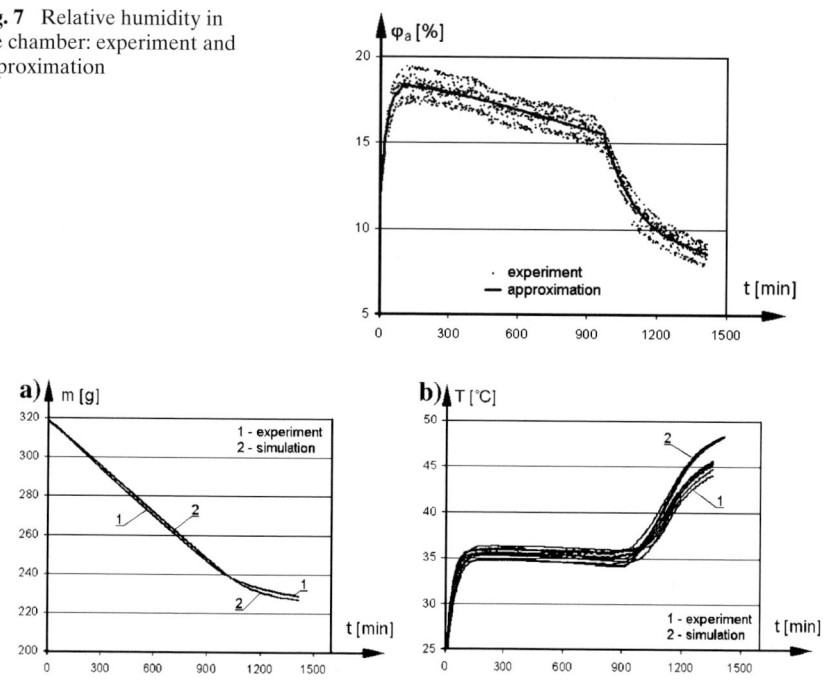

Fig. 7 Relative humidity in the chamber: experiment and approximation

Fig. 8 Comparison of the results of experiments and simulations—drying at 50.8°C: (**a**) change of sample mass, (**b**) temperature inside the sample

degree step on the basis of textbooks (Pikoń et al. 1972; Wiśniewski and Wiśniewski 1997; Strumiłło 1983).

The exchange coefficients $\alpha^{(m)}$ (Eq. 7) and $\alpha^{(\vartheta)}$ (Eq. 13) were estimated based on an experiment performed at the ambient temperature $T_a = 41.5°C$. It is assumed here that the coefficient of convective heat transfer $\alpha^{(\vartheta)}$ is constant and the coefficient of convective mass transfer $\alpha^{(m)}$ depends on ambient temperature and humidity. During the initial stage of the first period of drying the average ambient humidity was equal to $\varphi_a = 30.57\%$. The coefficient of convective mass transfer $\alpha^{(m)}$ in these drying conditions was calculated with the help of Eqs. 7–9 in which the mass flux j is taken from the experiment. Then the coefficient of convective heat transfer $\alpha^{(\vartheta)}$ was obtained from Eqs. 7, 12 and 13 together with the assumptions that the temperature of the dried body is equal to the wet bulb temperature and there is no heat flux inside the body. The coefficient of convective heat transfer $\alpha^{(\vartheta)}$ is treated as constant. Next, the values of the coefficient of convective mass transfer $\alpha^{(m)}$ in a wide range of the temperature and humidity of drying medium T_a, φ_a were calculated using Eqs. 7, 12, 13 together with the aforementioned assumptions. However, the temperature of the dried samples do not equal the wet bulb temperature. For that reason a new value of the coefficient of convective heat transfer, $\alpha^{(\vartheta)}$ was estimated using assumptions that the temperature of dried body is known from the experiment and there is no heat flux inside the body.

The numerical simulations of drying consist in solving the two-dimensional initial boundary problem (Eqs. 1–13), assuming homogeneous initial conditions and constant ambient temperature T_a. The evolution of the humidity φ_a in the dryer is

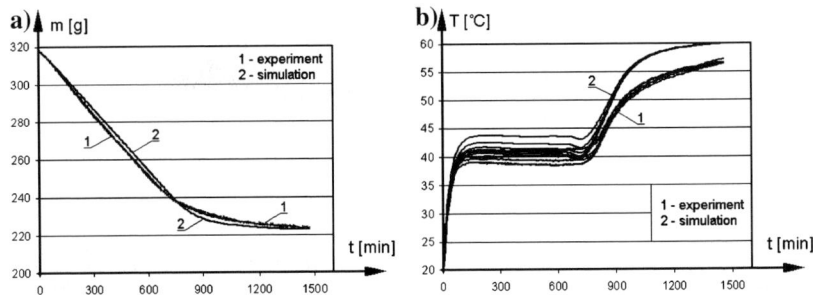

Fig. 9 Comparison of the results of the experiments and simulations—drying at 60.4°C: (**a**) change of sample mass, (**b**) temperature inside the sample

approximated with the continuous function (see Fig. 7), which is more convenient for the calculations than experimental points.

The problem is solved with the use of an explicit scheme of the final difference method. The spatial mesh consists of 101×51 points. The condition of stability indicates that the time step during the calculations should be less than 0.33 s. The time step taken in calculations is 0.01 s. During the second period of drying the moisture transfer coefficient Λ_X and the difference of the potentials $\Delta\mu$ are closely dependent on the temperature T and on the moisture content of the material surface $X|_{\partial\Omega^-}$ (see Fig. 2). The non-linearities could cause loss of the stability of the solution, so a special algorithm is adopted (see Musielak 2005).

The results of theoretical simulations together with experimental results are shown in Fig. 8 (ambient temperature 50.8°C) and Fig. 9 (ambient temperature 60.4°C). To compare these results it is necessary to introduce some measures. Therefore relative errors in the form

$$\delta_m(t) = \left| \frac{m_{\exp}(t) - m_{\text{sim}}(t)}{m_0 - m_r} \right|, \quad \delta_T(t) = \left| \frac{T_{\exp}(t) - T_{\text{sim}}(t)}{T_a - T_0} \right| \quad (13)$$

were introduced. In these errors, the differences between exprimental and simulation results are related to differences between initial and final stage of material.

Figures 8a and 9a show drying curves obtained numericaly and experimentally. The calculated curves are above the experimental ones during the first period of drying and below them in the second drying period. They show qualitative agreement between experiment and model. The average relative error δ_m is 1.6% and 2.3% for ambient temperature 50.8°C and 60.4°C, respectively. The maximum values of the error δ_m are 2.3% and 4.8%, respectively. It can thus be stated that the theoretical results differ slightly from the experimental data and quantitative agreement between them is obtained.

Figures 8b and 9b show the evolution of temperature during drying in chosen points (see Fig. 4). All curves show qualitative agreement between experiment and model, but the temperatures obtained theoretically are greater than the experimental values (all simulated curves are above the experimental ones). The relative error δ_T changes significantly during the process. Values of the error also depend on the measuring point. In the case of ambient temperature 50.8°C, the maximum values of the error δ_T are 4–9% during preheating, 0.5–2% during the first period of drying and 7–12% during the second period of drying. In the case of ambient temperature 60.4°C, the

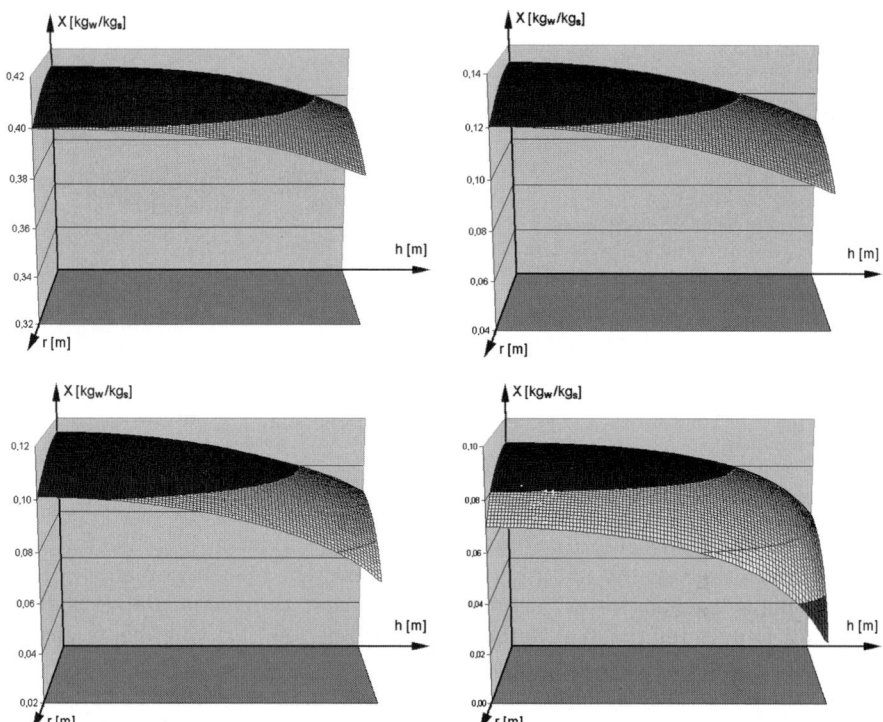

Fig. 10 Distributions of moisture content in dried cylinder after (**a**) 5 h of drying, (**b**) 11 h 20 min of drying, (**c**) 12 h of drying, (**d**) 13 h of drying

maximum values of the error δ_T are 3–7% during preheating and the first period of drying and 11–14% during the second period of drying. These values of the error δ_T show that quantitative agreement of temperature is not good, mainly because the material heat transfer properties used in calculations are taken from the literature, rather than being determined exactly for the kaolin used in the experiments. The fact that the simulated temperature is always greater than the experimental one testifies to this conclusion. It shows that improvement of estimation of heat transfer parameters should improve the agreement between simulated and experimental temperature.

Figure 10 shows how the moisture distribution in the kaolin sample is changed during the drying process. The kaolin sample (see Fig. 1) stands on the metal support and dries freely from the upper and side surfaces. Therefore, the most valuable fall of moisture content is observed at the point where the upper surface meets with the cylinder side (the point (R,H) in Fig. 1). The sample dries slowest in the center of the bottom surface.

During the first drying rate period (fig. 10a), gradients of moisture content are relatively small and do not change during the whole period. The beginning of the second drying rate period is noticed at about 11 h and 20 min. At this time the moisture content in the most dried point of the sample becomes lower than the critical moisture content $X_{kr} = 0.1$ (Fig. 10b). Transition from the first to the second drying rate period does not occur in the whole sample simultaneously but starts at every particular point of the surface when the moisture content at this point becomes lower than the critical

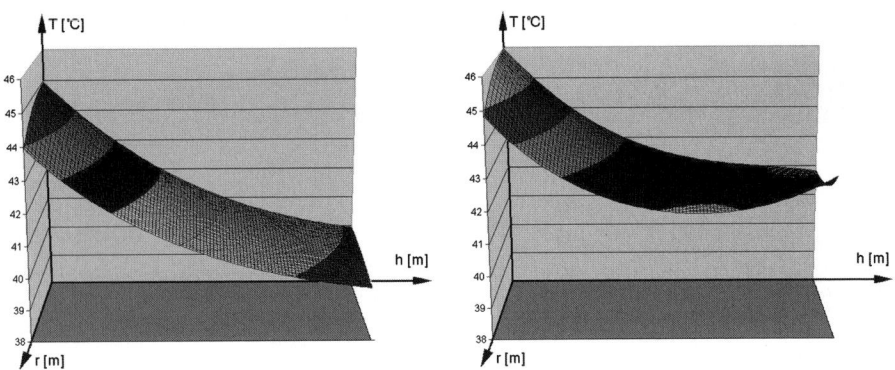

Fig. 11 Distributions of temperature after (**a**) 5 h of drying (**b**) 13 h of drying

one (Fig. 10c). Now the gradients of moisture content starts to rise, which is visible in the next two Figures (10c, d). Shortly after 12 h of experiment (Fig. 10c) the whole upper surface is below the critical moisture content. The moisture content in every part of sample becomes lower than the critical value during the next hour of drying (Fig. 10d). One can notice that during the second period of drying the gradient of moisture content is high, especially near the sample edges (Fig. 10c, d).

Such changes of moisture gradients result from decreasing moisture mass transport coefficient inside the sample when the moisture content is below the critical value, which means that less moisture is evaporated from the sample and the drying process becomes slower.

Figure 11 presents the temperature distribution inside the sample. After 5 h of drying (Fig. 11a) the sample is in the middle of the first drying period. The temperature is steady but different throughout the sample. Since there is no evaporation from the sample bedplate, the highest value of temperature is noticed there. The lowest temperature is observed on the upper surface of the dried cylinder. When the second drying period begins, the temperature inside the sample rises and the temperature distribution changes. A new extremum (local maximum) of temperature appears in the corner of the sample section (the point (R, H) in Fig. 1).

5 Conclusions

The main aim of the present work is validation of the model used with the help of experiment. In order to attain the validation, a number of material parameters need to be known. Parameters describing water, water vapour and humid air were taken from the literature. Parameters describing the skeleton and the constants describing diffusion of moisture inside the material were estimated on the basis of the literature and our own research. They were only estimated because there were no material parameters of KOC kaolin in the literature. The exchange coefficients specific for our set of laboratory dryer and kaolin sample were estimated based on experiment.

A detailed simulation of drying processes carried out at about 50°C and 60°C was then performed. The drying curves presented (Figs. 8a, 9a) show good agreement (both qualitative and quantitative) of the results obtained experimentally and those of numerical simulation. The simulation results well reflect the mass transfer, not only during the first period of drying but also during the second one and during the transition between these periods.

Curves of temperature evolution (Figs. 8b, 9b) show qualitative agreement between experiment and model. They show that the model could well reflect experimental temperature both during the first and the second periods of drying. Lack of quantitative agreement between experimental and theoretical temperature, especially during pre-heating and the second period of drying, results from estimation of material heat transfer properties used in calculations. These should be determinated exactly for the KOC kaolin used in experiments. Improvement in the estimation of heat transfer parameters should improve the agreement between simulated and experimental temperature.

The results described in this article show that the model presented well reflects moisture evaporation and temperature changes in the first and the second drying periods and is suitable for simulating unsteady drying processes.

Acknowledgements This work was carried out as a part of research project No 3T09C03028 sponsored by Ministry of Education and Science.

References

Augier, F., Coumans, W.J., Hugget, A., Kaasschieter, E.F.: On the study of cracking in clay drying. In: Proceedings of the 12th International Drying Symposium IDS2000, Noordwijkerhout, The Netherlands, 28-31 August 2000, Paper No.290 (2000)

Augier, F., Coumans, W.J., Hugget, A., Kaasschieter, E.F.: On the risk of cracking in clay drying. Chem. Eng. J. **86**, 133–138 (2002)

Banaszak, J.: Estimation of the effective coefficient of heat conduction for dried materials. In: Proceeding of the 11th Polish Drying Symposium XI PSS, Poznań, Poland, 13–16 September 2005, (in Polish)

Comini, G., Lewis, R.W.: A numerical solution of two-dimensional problems involving heat and mass transfer. Int. J. Heat Mass Transfer **19**, 1387–1392 (1976)

Coumans, W.J.: Models for drying kinetics based on drying curves of slabs. Chem. Eng. Proc. **39**, 53–68 (2000)

Ketelaars, A.A.J.: Drying Deformable Media. Kinetics, Shrinkage and Stresses. PhD Thesis, Technische Universiteit Eindhoven (1992)

Kowalski S.J.: Drying processes involving permanent deformations of dried materials. Int. J. Eng. Sci. **34**(13), 1491–1506 (1996)

Kowalski, S.J.: Thermomechanics of Drying Processes. Springer, Berlin (2003)

Kowalski, S.J., Strumiłło, Cz.: Moisture transport in dried materials. Boundary conditions. Chem. Eng. Sci. **40**(7), 1141–1150 (1997)

Luikov A.V.: Systems of differential equations of heat and mass transfer in capillary porous bodies (Review). Int. J. Heat and Mass Transfer **18**(1), 1–14 (1975)

Musielak, G.: Influence of the drying medium parameters on drying induced stresses. Dry. Technol. **18**(3), 561–581 (2000)

Musielak, G.: Modelling and Numerical Simulation of the Transport Phenomena and the Drying Induced Stresses inside of the Capillary-Porous Materials. Wydawnictwo Politechniki Poznańskiej, seria Rozprawy, nr 386 (in Polish) (2004a)

Musielak, G.: Modelling of the heat and mass transport phenomena in kaolin during the first and the second periods of drying. Chem. Proc. Eng. **25**, 393–409, (in Polish) (2004b)

Musielak, G.: Numerical treatment of non-linear boundary condition of mass transfer during second period of drying. In: Proceeding of the 11th Polish Drying Symposium XI PSS, Poznań, Poland, 13–16 September 2005, (in Polish)

Musielak, G., Banaszak, J., Kasperek, J.: Determination of temperature dependence of diffusion coefficient in fully saturated kaolin. In: Proceeding of the 11th Polish Drying Symposium XI PSS, Poznań, Poland, 13–16 September 2005, (in Polish)

Pabis St.: Theoretical models of vegetable drying by convection. In: Proceeding Of The 11th Polish Drying Symposium XI PSS, Poznań, Poland, 13–16 September 2005

Philip J.R., De Vries D.A.: Moisture movement in porous materials under temperature gradients. Trans. A. Geophys. Un. **38**(2), 222–232 (1957)

Pikoń, J., Hehlmann, J., Sąsiadek, B., Janowicz, R.: Chemical Apparatus. Wydawnictwo Politechniki Śląskiej, Gliwice, (in Polish) (1972)

Strumiłło, Cz.: Foundations of Theory and Technology of Drying. 2nd edn. Wydawnictwo Naukowo Techniczne Warszawa, (in Polish) (1983)

Whitaker S.: Simultaneous heat, mass and momentum transfer in porous media: a theory of drying. Adv. Heat Mass Transfer **13**, 110–203 (1977)

Wiśniewski, S., Wiśniewski, T.S.: Heat Exchange, 2nd edn. Wydawnictwo Naukowo Techniczne Warszawa, (in Polish) (1997)

http://www.surmin-kaolin.com.pl/english/products/ceramic.html

Effects of the method of identification of the diffusion coefficient on accuracy of modeling bound water transfer in wood

Wieslaw Olek · Jerzy Weres

Received: 10 November 2005 / Accepted: 26 March 2006 /
Published online: 29 September 2006
© Springer Science+Business Media B.V. 2006

Abstract An alternative approach to determining the bound water diffusion coefficient is proposed. It comprises a method for solving the inverse diffusion problem, an improved algorithm for the bound-constrained optimization as well as an alternative submodel for the diffusion coefficient's dependency on the bound water content. Identification of the diffusion coefficient for Scots pine wood (*Pinus sylvestris* L.) using the proposed inverse approach is presented. The accuracy of predicting the diffusion process with the use of the coefficient values determined by traditional sorption methods as well as by the inverse modeling approach is quantified. The similarity approach is used and the local and global relative errors are calculated. The results show that the inverse method provides valuable data on the bound water diffusion coefficient as well as on the boundary condition. The results of the identification can significantly improve the accuracy of mass transfer modeling as studied for drying processes in wood.

Keywords Inverse method · Computer-aided identification · Fick's law · Scots pine wood · Sorption experiments · Validation of identification

Nomenclature
a Constant in Eqs. 13 and 14 (−)
b Constant in Eq. 14 (−)
D Diffusion coefficient (m^2/s)
D_0 Coefficient in Eqs. 12–14 (m^2/s)

W. Olek (✉)
Department of Mechanical Engineering and Thermal Techniques,
Agricultural University of Poznań, ul. Wojska Polskiego 28,
60-637 Poznań, Poland
e-mail: olek@au.poznan.pl

J. Weres
Institute of Agricultural Engineering,
Agricultural University of Poznań, ul. Wojska Polskiego 28,
60-637 Poznań, Poland
e-mail: weres@au.poznan.pl

e_1	Local relative error (%)
e_2	Global relative error (%)
E	Reduced bound water content (−)
l	Half-thickness (m)
M	Water content (kg/kg$_{\text{dry base}}$)
M_∞	Equilibrium water content (kg/kg$_{\text{dry base}}$)
M_0	Initial water content (kg/kg$_{\text{dry base}}$)
$M(t)$	Global water content in selected time instants (kg/kg$_{\text{dry base}}$)
$M_{\exp}(t)$	Experimental water content (kg/kg$_{\text{dry base}}$)
$M_{\text{pred}}(t)$	Predicted water content (kg/kg$_{\text{dry base}}$)
NT	Number of time intervals in computation
$NT_{\exp}$	Number of time intervals in experiment
S	Objective function
t	Time (s)
t_F	Final time (s)
w_i	Weight function
x	Space dimension (m)

Greek symbols

Γ	Points located at the two boundary sides of the domain (two points in the one-dimensional model)
σ	Surface emission coefficient (m/s)
Ω	Geometric domain of the R^1 space
$\overline{\Omega}$	Geometric domain of the R^1 space with the boundary

1 Introduction

The modeling of wood drying as well as the description of moisture transfer in products made of wood already dried below the fiber saturation point requires credible data on the material properties. The bound water diffusion coefficient describes one of the most important properties responsible for obtaining good quality predictions of water transfer processes. The diffusion coefficient is usually determined using the sorption method. During recent decades a number of different approaches to the sorption method have been developed and applied. The simplest and the most frequently used was the initial sorption technique (e.g., Siau 1995; Stamm 1959, 1960; Wadsö 1994). Unfortunately, the assumptions that the coefficient does not depend on moisture content and the wood surface comes instantly to hygroscopic equilibrium with moist air were the most important disadvantages of the technique. It was shown early on that the second assumption is not valid in the case of wood (Skaar 1988; Shmulsky et al. 2002). Therefore, methods were developed, which took into consideration the gradual approach of the wood surface to equilibrium. The methods were able to separate the surface resistance from the internal resistance by identifying both the surface emission coefficient and the bound water diffusion coefficient in wood (Choong and Skaar 1969, 1972). Unfortunately, the methods were based on the analytical solution of the diffusion problem, which required the assumption of a constant value of the coefficient. This may cause significant errors in the identification of the coefficients and lead to false values. The problem of false values was extensively reported by Söderström and Salin (1993).

The inaccuracy in the diffusion coefficient's determination may be significantly reduced by applying the inverse approach for solving mass transport problems. The authors of the present paper have already shown the effectiveness of the inverse problem concept for analyzing the heat and mass transport in forest products (Weres et al. 2000; Olek et al. 2005; Weres and Olek 2005). However, the improvement in the accuracy of the diffusion coefficient's identification obtained by the application of the inverse approach was never quantified. Therefore, the objective of the present paper was to analyze the accuracy of predicting the diffusion process with the use of the coefficient values determined by traditional sorption methods as well as the inverse modeling approach.

2 Methods

2.1 Experimental data

Experimental material was obtained from freshly cut Scots pinewood (*Pinus silvestris* L.) with clearly oriented growth rings and uniform density of $490 \pm 10 \, \text{kg/m}^3$. The material was carefully dried to a moisture content lower than the fiber saturation point and then equilibrated to an initial moisture content of ca. $0.08 \, \text{kg/kg}_{\text{dry base}}$. Two types of samples, i.e., radial and tangential ones, in the shape of rectangular prisms, were cut from the equilibrated material. The dimensions of the samples were 70, 35 and 8 mm. The last dimension always corresponded to the thickness ($2l$) of a sample. The sides of the samples were sealed by covering with a few layers of chlorinated rubber enamel (Fig. 1).

The sorption experiments were performed in a setup with controlled air parameters, i.e., temperature of 23.4°C, relative humidity corresponding to the equilibrium moisture content of ca. $0.14 \, \text{kg/kg}_{\text{dry base}}$ and forced air flow of 0.8 m/s. The individual samples were placed on a balance of accuracy of 0.001 g and weighed at time instants. The results were collected by a data acquisition system. The final moisture content of each sample was determined by the gravimetric method after the sorption experiment. The results of sorption experiments were stored in an empirical database as the moisture content values of the samples depending on the time instants.

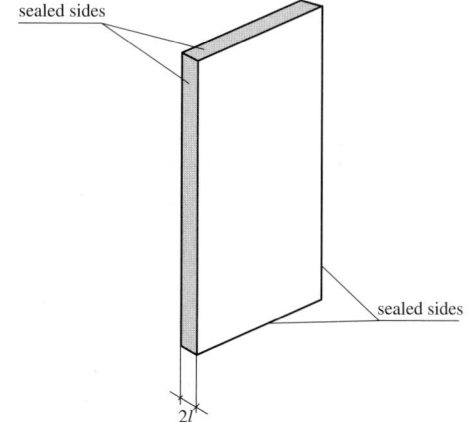

Fig. 1 Schematic representation of a sample with four sealed sides

2.2 Mathematical model

The mathematical model of the unsteady state bound water diffusion is traditionally given by the transient form of Fick's second law:

$$\frac{\partial M}{\partial t} = \frac{\partial}{\partial x}\left(D\frac{\partial M}{\partial x}\right), \quad (x,t) \in \Omega \times [0, t_F] \tag{1}$$

with the initial condition

$$M(x, 0) = M_0, \quad x \in \overline{\Omega} \tag{2}$$

and the third kind boundary condition:

$$\left(-D\frac{\partial M}{\partial x}\right) = \sigma \cdot [M(x,t) - M_\infty], \quad (x,t) \in \Gamma \times [0, t_F], \tag{3}$$

$$\left(-D\frac{\partial M}{\partial x}\right) = \sigma \cdot [M(x,t) - M_\infty] \quad (x,t) \in \Gamma \times [0, t_F], \tag{4}$$

or the first kind boundary condition:

$$M(x, t) = M_\infty, \quad (x,t) \in \Gamma \times [0, t_F], \tag{5}$$

$$M(x, t) = M_\infty, \quad (x,t) \in \Gamma \times [0, t_F]. \tag{6}$$

2.3 Methods used for diffusion coefficient estimation

The diffusion coefficient was estimated with the use of the two traditional sorption methods as well as the inverse approach. The basic assumptions of all methods applied in the present analysis are given below.

2.3.1 Initial sorption

The diffusion problem given by (1), (2), (5), and (6) has an analytical solution for the initial stage of sorption with the assumption of a constant value of the diffusion coefficient. The solution can be written for the total bound water content and reduced to the following relation (Crank 1975):

$$E = 2\left(\frac{Dt}{\pi l^2}\right)^{1/2}, \tag{7}$$

where for adsorption, reduced bound water content (E) is defined as

$$E(t) = \frac{M(t) - M_0}{M_\infty - M_0}. \tag{8}$$

The diffusion coefficient was calculated from results of the sorption experiments presented in the form of plots of E versus $\sqrt{t}$ (linear part of the relation) and the following equation derived from (7):

$$D = \frac{\pi}{4/l^2}\left(\frac{\Delta E}{\Delta \sqrt{t}}\right)^2. \tag{9}$$

2.3.2 Separation of diffusion and surface emission coefficients

Liu (1989) proposed a method for the simultaneous determination of the diffusion coefficient as well as surface emission coefficient. It was based on the analytical solution of the diffusion problem given by (1)–(4) and originally obtained by Newman (1931). The earlier methods of coefficient separation (Choong and Skaar 1969, 1972) required the results of sorption experiments done on samples of different thickness. Liu (1989) proposed an analytical procedure that required results to be obtained for one thickness only. The diffusion coefficient was determined from the relation:

$$D = l^2 \frac{-0.1654}{0.7010 \cdot dt/dE + 2.05 \cdot t}\bigg|_{E=0.5} \quad (10)$$

and the surface emission coefficient (σ) was calculated as:

$$\sigma = \frac{0.7010 \cdot D}{D \cdot t/l - 0.1963 \cdot l}\bigg|_{E=0.5}. \quad (11)$$

2.3.3 Inverse method

The inverse finite element analysis approach to identifying the diffusion coefficient was based on the inverse problem concept, optimization techniques, and the finite element method for solving direct problems. A detailed description of the approach was given by Olek et al. (2005) and Weres and Olek (2005). The operational form of the mathematical model of the analyzed diffusion process was developed by applying the finite element approximation (isoparametric elements in space and an absolutely stable, threepoint recurrence scheme in time), the iteration procedure resulting from the quasi-linearity of the analyzed diffusion model, and the empirical submodels for the diffusion coefficient. The empirical submodels analyzed in the present study were parameterized by the following functions:

$$D = D_0, \quad (12)$$

$$D = D_0 \cdot \exp[-a \cdot M(t)], \quad (13)$$

$$D = D_0 \cdot \exp\left[-\left(a \cdot M(t) + b \cdot M^2(t)\right)\right], \quad (14)$$

where D_0, a, and b are the estimated coefficients. The bound water content at the hygroscopic equilibrium (M_∞) was taken as an additional parameter, estimated in all options of the inverse analysis.

The algorithm for solving the direct diffusion problem was supplemented with the module for averaging all predicted nodal values of the bound water content in the geometric domain. This allowed us to obtain the global values of the water content $M_{\text{pred}}(t)$ for each time instant. The predicted global values were used in the procedure of coefficient estimation.

The optimization algorithm was developed on the basis of the trust regions and the variable metric approach, in which gradients and Jacobians were approximated automatically, and Hessians were approximated by the quasi-Newtonian technique. The objective function was defined as the sum of the squares of the residuals of the measured and predicted global values of the bound water content (Eq. 15). We assumed $w_i = 1$, as a result of our previous analyses of differentiating values of the weight function. The invented inverse problem approach was implemented and the software package was coded in Lahey/Fujitsu Fortran 95.

$$S = \sum_{i=1}^{NT_{\exp}} w_i \left[M_{\exp}(t_i) - M_{\text{pred}}(t_i) \right]^2. \tag{15}$$

3 Results

The diffusion coefficient was identified for Scots pine wood in the radial as well as tangential directions. The input data consisted of the averaged results of adsorption experiments (at least results of four processes were subjected to averaging). The values of the diffusion coefficient as well as the surface emission coefficient determined with the initial sorption method and the Liu method are presented in Table 1, and the values of the coefficients determined with the inverse method are shown in Table 2. The latter values of the coefficients were used in modeling the diffusion process. The numerically obtained bound water content changes in time were compared to the empirical data of experiments that were not used for the identification (Figs. 2, 3). In order to quantify the quality of the coefficient estimated by the three analyzed methods, two errors were defined and calculated (Olek et al. 2003, 2005), i.e., the local relative error (e_1)

$$e_1(t_i) = 100 \frac{|M_{\exp}(t_i) - M_{\text{pred}}(t_i)|}{M_{\exp}(t_i)}, \quad i = 1, \ldots, NT_{\exp} \tag{16}$$

Table 1 Values of the coefficients estimated by the analytical sorption methods

	Applied method	σ (m/s)	D (m^2/s)
Radial direction	Initial sorption	–	$1.105 \cdot 10^{-10}$
	Liu method	$2.343 \cdot 10^{-7}$	$1.471 \cdot 10^{-10}$
Tangential direction	Initial sorption	–	$0.6794 \cdot 10^{-10}$
	Liu method	$1.218 \cdot 10^{-7}$	$0.9592 \cdot 10^{-10}$

Table 2 Values of the coefficients estimated by the inverse method

	Inverse method option	σ (m/s)	D_0 (m^2/s)	$a(-)$	$b(-)$	M_∞ (kg/kg$_{\text{drybase}}$)	S
Radial direction	Eq. 12	$6.751 \cdot 10^{-7}$	$1.707 \cdot 10^{-10}$	–	–	0.1376	0.09242
	Eq. 13	$5.515 \cdot 10^{-7}$	$2.714 \cdot 10^{-10}$	2.400	–	0.1377	0.08928
	Eq. 14	$6.752 \cdot 10^{-7}$	$0.006262 \cdot 10^{-10}$	−78.80	288.0	0.1378	0.08460
Tangential direction	Eq. 12	$2.081 \cdot 10^{-7}$	$1.042 \cdot 10^{-10}$	–	–	0.1436	0.05195
	Eq. 13	$3.031 \cdot 10^{-7}$	$0.3041 \cdot 10^{-10}$	−6.782	–	0.1433	0.02541
	Eq. 14	$3.416 \cdot 10^{-7}$	$0.005259 \cdot 10^{-10}$	−60.54	185.0	0.1433	0.02442

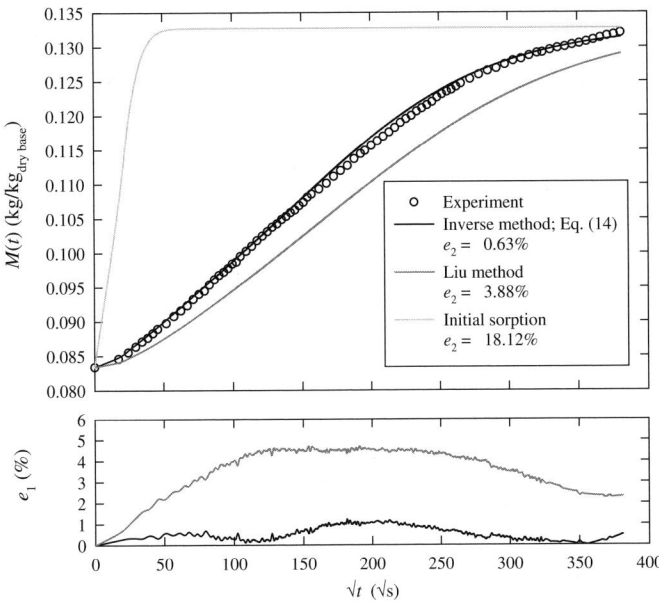

Fig. 2 Validation of the diffusion coefficient estimation, Scots pine, and radial direction

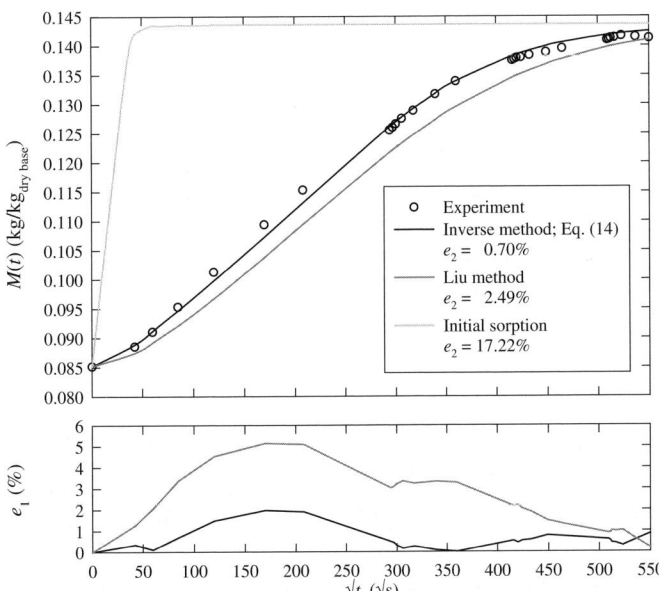

Fig. 3 Validation of the diffusion coefficient estimation, Scots pine, and tangential direction

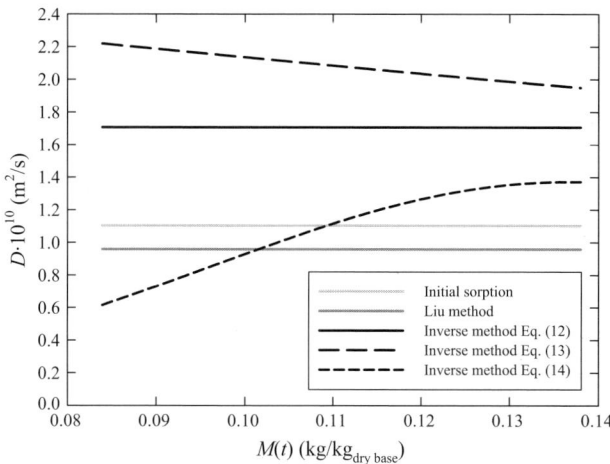

Fig. 4 Diffusion coefficient changes with moisture content, Scots pine, and radial direction

and the global relative error (e_2)

$$e_2 = 100 \frac{\sqrt{\sum_{i=1}^{NT_{\exp}} \left[M_{\exp}(t_i) - M_{\text{pred}}(t_i)\right]^2}}{\sqrt{\sum_{i=1}^{NT_{\exp}} M_{\exp}(t_i)^2}}. \qquad (17)$$

For clarity, the plots (Figs. 2, 3) of the predicted values of the water content as well as the values of the local relative error (e_1) are presented for only one set of results of the inverse method identification, for which the objective function (S) had the smallest values. In both cases, the inverse identification was made for the diffusion coefficient parameterization given by Eq. 14, i.e., the three parameter submodel (Table 2). Figures 2 and 3 do not contain the values of the local relative error obtained for the diffusion prediction obtained for the coefficients identified with the use of the initial sorption method since the maximum values of the error were as high as 60%.

For the analyzed sorption processes and the empirical submodel given by Eq. 14 the obtained values of the diffusion coefficient were augmented with the increase of the bound water content. This supports the common opinion that the diffusion coefficient dependens on the bound water content (e.g., Skaar 1988). However, the results presented in Table 2 show that in case of the simplest submodel (Eq. 13) the dependency may be opposite (e.g., radial direction). Therefore, the proper selection of the empirical function describing the dependency on the water content has great importance. The changes of the diffusion coefficient values with the moisture content for all options of the identification are presented in Fig. 4 (radial direction) and Fig. 5 (tangential direction), respectively.

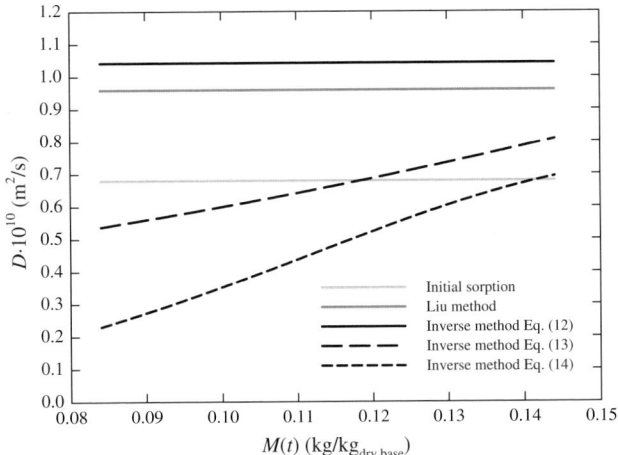

Fig. 5 Diffusion coefficient changes with moisture content, Scots pine, and tangential direction

4 Conclusions

The analysis of similarity between the diffusion modeling and the results of experiments allowed us to derive the following conclusions:

(1) Application of the initial sorption method is not acceptable from the point of view of diffusion coefficient identification in wood. The local relative error values reached as much as 60%. The assumption of the first kind boundary condition is not valid in the case of wood.
(2) The application of the analytical method which assumes the convective boundary condition, does not significantly improve the accuracy of the diffusion identification (global relative error of almost 20%). This was primarily due to the assumption of constant values of the diffusion coefficient as well as the experimental determination of the bound water content at the hygroscopic equilibrium.
(3) The best similarity was obtained for the results of the inverse identification with use of the submodel given by Eq. 14. However, a further improvement can be achieved by taking into consideration the coupling between diffusion and the heat transfer, as well as by modifying the submodel describing the diffusion coefficient's dependency on the water content.

References

Choong, E. T., Skaar, C.: Separating internal and external resistance to moisture removal in wood drying. Wood Sci. **1**, 200–202 (1969)
Choong, E. T., Skaar, C.: Diffusivity and surface emissivity in wood drying. Wood Fiber **4**, 80–86 (1972)
Crank, J.: The Mathematics of Diffusion. Clarendon Press, Oxford (1975)
Liu, J. Y.: A new method for separating diffusion coefficient and surface emission coefficient. Wood Fiber Sci. **21**, 133–141 (1989)
Newman, A. B.: The drying of porous solids: diffusion and surface emission equations. Trans. AIChE **27**, 203–220 (1931)
Olek, W., Weres, J., Guzenda, R.: Effects of thermal conductivity data on accuracy of modeling heat transfer in wood. Holzforschung **57**, 317–325 (2003)

Olek, W., Perré, P., Weres, J.: Inverse analysis of the transient bound water diffusion in wood. Holzforschung **59**, 38–45 (2005)

Shmulsky, R., Kadir, K., Erickson, R.: Effect of air velocity on surface EMC in the drying of red oak lumber. For. Prod. J. **52**, 78–80 (2002)

Siau, J.F.: Wood: Influence of Moisture on Physical Properties. Virginia Tech, Blacksburg (1995)

Skaar, C.: Wood-water Relations. Springer-Verlag, Berlin (1988)

Söderström, O., Salin J. G.: On determination of surface emission factors in wood drying. Holzforschung **47**, 391–397 (1993)

Stamm, A.J.: Bound water diffusion into wood in the fiber direction. For. Prod. J. **9**, 27–32 (1959)

Stamm, A. J.: Bound water diffusion into wood in the across the fiber direction. For. Prod. J. **10**, 524–528 (1960)

Wadsö, L.: A test of different methods to evaluate the diffusivity form sorption measurements. Drying Techn. **12**, 1863–1876 (1994)

Weres, J., Olek, W., Guzenda, R.: Identification of mathematical model coefficients in the analysis of the heat and mass transport in wood. Drying Technol. **18**, 1697–1708 (2000)

Weres, J., Olek, W.: Inverse finite element analysis of technological processes of heat and mass transport in agricultural and forest products. Drying Technol. **23**, 1737–1750 (2005)

Stresses in dried wood. Modelling and experimental identification

Stefan Jan Kowalski · Anna Smoczkiewicz-Wojciechowska

Received: 8 December 2005 / Accepted: 26 March 2006 /
Published online: 28 September 2006
© Springer Science+Business B. V. 2006

Abstract The paper presents a simple mathematical model of drying that permits evaluation of moisture content distribution in dried wood during the constant and falling drying rate periods and, in particular, estimation of stresses generated from the moment when the moisture content at the body surface reaches the fibre saturation point (FSP). The acoustic emission method (AE) is used for monitoring the state of stress in dried wood. The numerically evaluated drying induced stresses are compared with the number of acoustic signals and their energy monitored on line during drying tests. It can be stated that the enhanced emission of acoustic signals occurs at those moments when the drying induced stresses approach their maximum. Both the numerical calculus and the experimental tests were conducted on a pine-wood sample in the form of a disk.

Keywords Shrinkage · Stresses · Destruction · Modelling · Acoustic emission

Nomenclature

A	Elastic bulk modulus [MPa]
$B = kR/\Lambda$	Coefficient of mass exchange (Biot number) [1]
D	Coefficient of diffusion [m²/s]
e_{ij}	Strain deviator [1]
J_0, J_1	Bessel functions of first kind of zero and first order
k	Coefficient of convective vapour exchange [kg/ m²·s]
M	Elastic shear modulus [MPa]
r, R	Cylinder radius [m]
s_{ij}	Stress deviator [Pa]
t	Time [s]

S. J. Kowalski (✉)· A. Smoczkiewicz-Wojciechowska
Institute of Technology and Chemical Engineering, Poznań University of Technology, pl. Marii Skłodowskiej Curie 2, 60-965 Poznań, Poland
e-mail: stefan.j.kowalski@put.poznan.pl

T	Temperature [K]
u	Radial displacement vector [m]
W	Mass flux of moisture [kg/m^2·s]
X	Dry basis moisture content [1]
Y	Vapour content in drying air [1]

Greek symbols

α, β	Ratios of mechanical modules [1]
κ	Viscous bulk modulus [Pa·s]
$\kappa^{(T)}$	Coefficient of thermal expansion [1/K]
$\kappa^{(X)}$	Coefficient of humid expansion [1]
ε_{ij}	Strain tensor [1]
ε	Volumetric strain [1]
λ_n, α_n	Eigenvalues [1]
σ_{ij}	Stress tensor [Pa]
σ	Spherical stress [Pa]
ρ	Mass density [kg/m^3]
η	Shear viscoelastic modulus [Pa·s]
Ω, ω	Parameters [1]
$\vartheta = TT_r$	Relative temperature [°C]
$\theta = X - X_r$	Relative moisture content [1]
τ	Retardation time [s]
Λ	Mass transport coefficient [kg·s/m^3]

1 Introduction

It is known that wood is a material that is sensitive to drying because of its tendency to crack during the process. The main reason for crack formation is the stresses generated due to the non-uniform shrinkage of wood. As has been stated in the literature (see e.g. Glijer, et al. 1984; Pang, 2000; Kowalski, et al. 2004), wood suffers the greatest shrinkage and has the weakest strength in the direction tangential to annual rings. Therefore, wood cracking proceeds mostly in the direction perpendicular to the tangents to annual rings.

In order to protect wood against destruction during drying one ought to study deeply the factors that promote the development of stresses. The best method for such a study is to construct a possibly precise drying model that enables numerical analysis of stresses in wood and their time evolution. Such a model usually contains a number of coefficients referring to both the dried material and the drying medium. The parameters of drying medium (air) permit control of drying processes in this way minimizing the negative effects caused by the drying induced stresses.

The aim of this paper is to present a simplified drying model based on the diffusion equation with different coefficients and boundary conditions for the first and second period of drying. Such a model enables estimation of moisture content distribution in the dried sample in several instants of time and in this way also the determination of stresses, the magnitude of which depends inter alia on the moisture distribution gradient.

Describing stresses in dried wood, we take into account the fact that wood shrinks below the so-called *fibre saturation point* (FSP), which means that wood is stress free

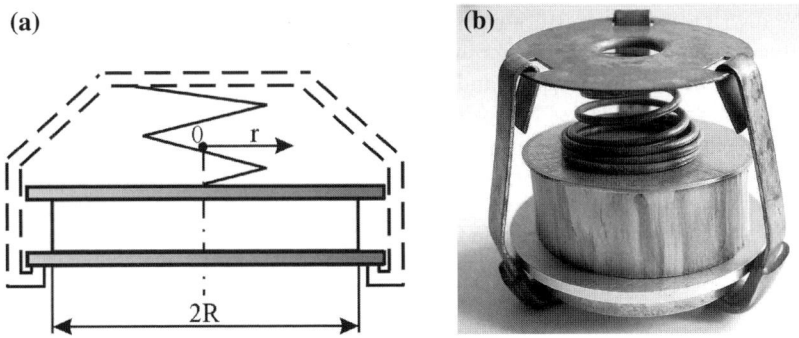

Fig. 1 Pinewood disk placed in dying holder: (**a**) diagram, (**b**) photograph

when the moisture content is over FSP. Thus, the shrinkage and the stresses occur first at the surface of the sample when the moisture content approaches the FSP. After that, as wood dries, the drier surface attempts to shrink but is restrained by the wet core. The surface is stressed in tension and the core in compression.

In our experimental studies, the development of the drying induced stresses in wood is monitored with the help of the acoustic emission method (AE) (see Kowalski and Smoczkiewicz, 2004; Kowalski, et al. 2004), which permits *on line* monitoring of the number of acoustic signals and their energy emitted by the cracking wood structure. We seek to show that the enhanced AE begins at that time when the drying induced stresses start to grow. In this way, we can experimentally detect the moments when the stresses tend to their maximum.

In this paper, both the numerical calculus and the stresses monitored experimentally with the help of AE method are performed on a pine-wood disk dried convectively.

2 Simplified mathematical model of drying

A rigorous thermomechanical model of drying constructed for the aim of describing mechanical effects during drying (deformation, stresses) has been presented elsewhere (see e.g. Kowalski, 2003). In this paper we present an analytical solution of differential equations describing the moisture content and the stress distribution in a pine-wood disk. Therefore, a number of simplifications with respect to the rigorous model are introduced. These are expressed by the following assumptions:

- The sample under drying is a cylindrical disk of radius R, insulated against moisture removal on the upper and lower surfaces (see Fig. 1). Thus, drying proceeds only through the lateral surface. The numerical problem of such geometry can be considered as cylindrically symmetric.
- The influence of temperature on shrinkage is neglected as being much smaller than the shrinkage owing to moisture removal, so that the non-linear moisture content distribution is the only reason for disk deformation and stress generation.
- The influence of wood deformation on the transport of moisture is neglected. Similarly, the thermodiffusional transport of moisture is considered insignificant.

- The distribution of moisture content in the disk sample is described by a linear diffusion equation with constant (averaged) diffusion coefficients, which, however, are different for the saturated and the unsaturated regions of wood.
- The disk sample is stress free when the moisture content exceeds the FSP. The shrinkage occurs and the stresses arise when the moisture content in some area of the disk drops below the FSP.
- Wood is assumed to exibit rheological properties described by the Maxwell physical relation. The physical relation for a woody disk is similar to that for an isotropic body but with averaged material constants in both tangential and radial directions.

In order to construct the differential equation describing the distribution of moisture content in a dried disk, the following mass balance equations for the solid skeleton (wood fibres) and the moisture in wood pores are needed

$$\dot{\rho}_s + \rho_s \text{div}\boldsymbol{u} = 0, \qquad \dot{\rho}_m + \rho_m \text{div}\boldsymbol{u} = -\text{div}\boldsymbol{W}, \tag{1}$$

where ρ_s and ρ_m denote the mass concentrations of the solid skeleton and the moisture (partial mass densities), $\boldsymbol{u}$ is the displacement vector of the solid skeleton, and $\boldsymbol{W}$ is the moisture flux (amount of moisture per unit area and unit time). A dot over a symbol denotes the time derivative.

Introducing the dry basis moisture content defined as $X = \rho_m/\rho_s$, we can rewrite the mass balance equation for moisture content as follows

$$\rho_s \dot{X} = -\text{div}\boldsymbol{W}, \tag{2}$$

Based on the thermodynamic inequality, it was concluded (see Kowalski, 2003) that the moisture flux is proportional to the gradient of moisture potential, which is a function of the state parameters, such as temperature, strain of the body, and moisture content. In light of the above assumptions, we state that the gradient of moisture content is the main force responsible for moisture transport in the present considerations, that is

$$\boldsymbol{W} = -\Lambda \text{grad} X, \tag{3}$$

where Λ is termed the moisture transport coefficient.

Substituting the rate Eq. (3) into the equation of mass continuity for moisture (2), we obtain the diffusion equation describing the moisture content distribution in a dried body as a function of time

$$\dot{X} = D\nabla^2 X \quad \text{with} \quad D = \Lambda/\rho_s \tag{4}$$

The value of the diffusion coefficient D will be assumed different for the saturated and the unsaturated wood.

Wood has an anisotropic structure (a kind of orthotropy). Considering the thin woody disk in this paper, we shall neglect its anisotropic structure. Attention is concentrated on the saturated and the unsaturated regions of the disk (see Fig. 2), which differ from each other in mechanical properties. As it has already been stated, see e.g. Kowalski et al. (2004), the mechanical properties of wood depend on moisture content in the range from dry state to the FSP. Over the FSP these properties become constant and no shrinkage occurs.

In this paper, we analyse the drying induced stresses in a woody disk twice: first, assuming that the material is elastic, and second, that it is viscoelastic. The elastic material is described by physical relations of the form Kowalski (2003)

$$s_{ij} = 2M e_{ij}, \quad \sigma = K(\varepsilon - \varepsilon^{(TX)}) \tag{5}$$

Fig. 2 Pinewood disk with separated saturated (S) and unsaturated (US) regions

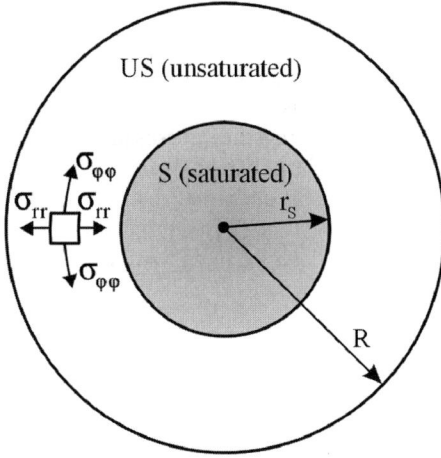

The viscoelastic material (Maxwell model) is expressed by the following physical relations

$$\dot{s}_{ij} + \frac{M}{\eta}s_{ij} = 2M\dot{e}_{ij}, \qquad \dot{\sigma} + \frac{K}{\kappa}\sigma = K(\dot{\varepsilon} - \dot{\varepsilon}^{(TX)}) \qquad (6)$$

In the above relations, $s_{ij} = \sigma_{ij} - \sigma\delta_{ij}$ is the stress deviator, $\sigma = \sigma_{ii}/3$ is the spherical part of stress, $e_{ij} = \varepsilon_{ij} - (\varepsilon/3)\delta_{ij}$ is the strain deviator, $\varepsilon = \varepsilon_{ii}$ is the volumetric strain, M and K are the elastic shear and bulk moduli, and η and κ are the viscoelastic shear and bulk moduli. The material moduli generally depend on the moisture content, but they are constant for wood over the FSP. In the present considerations, they are also taken as constants averaged from minimum and maximum values in the range below the FSP, that is from zero moisture content to the FSP.

The temperature and moisture content involve the volumetric thermal–humid strain

$$\varepsilon^{(TX)} = 3(\kappa^{(T)}\vartheta + \kappa^{(X)}\theta), \qquad (7)$$

where $\vartheta = T - T_\mathrm{r}, \theta = X - X_\mathrm{r}$, with T_r and X_r as the reference temperature and moisture content. According to the above assumption, we neglect here the volumetric strain caused by temperature, i.e., we take the coefficient of thermal expansion as $\kappa^{(T)} \approx 0$. The coefficient of humid expansion fulfils the following criterion

$$\kappa^{(X)} = \begin{cases} 0 & \text{for } \theta_\mathrm{s} \le \theta \le \theta_0, \\ \kappa_0^{(X)} & \text{for } \theta_\mathrm{e} \le \theta \le \theta_\mathrm{s}, \end{cases}$$

where $\theta_0, \theta_\mathrm{s}$ and θ_e denote the initial moisture content, the fibre saturation point, and the final (equilibrium) moisture content, respectively.

The stresses in viscoelastic material (v) can be expressed as the stresses calculated for elastic material (e) using the Borel convolution formula (see Kowalski and Rajewska, 2002)

$$\sigma_{ij}^{(v)}(r,t) = \sigma_{ij}^{(e)}(r,t) - \frac{1}{\tau}\int_0^t \exp\left(-\frac{t-\xi}{\tau}\right)\sigma_{ij}^{(e)}(r,\xi)\mathrm{d}\xi, \qquad (8)$$

where τ denotes the relaxation time.

3 Distribution of moisture content

3.1 First (I) period of drying

The distribution of moisture content in a pine-wood disk during the first period of drying, also called the constant drying rate period, is described by the following differential equation and boundary and initial conditions

$$\frac{\partial \theta^{(I)}}{\partial t} = D^{(I)} \left(\frac{\partial^2 \theta^{(I)}}{\partial r^2} + \frac{1}{r} \frac{\partial \theta^{(I)}}{\partial r} \right), \tag{9}$$

$$\frac{\partial \theta^{(I)}}{\partial r}\bigg|_{r=0} = 0, \quad -\Lambda^{(I)} \frac{\partial \theta^{(I)}}{\partial r}\bigg|_{r=R} = k^{(I)} (Y_n - Y_a) = \text{const}, \tag{10}$$

$$\theta^{(I)}(r, t)\big|_{t=0} = \theta_0 = \text{const}, \tag{11}$$

where Y_n and Y_a are the vapour contents in drying air close to and far from the disk surface, θ_0 is the initial moisture content in the disk, and $k^{(I)}$ is the coefficient of convective vapour exchange between the dried disk and the ambient air, respectively.

The boundary condition (10) on the left express the symmetry of moisture distribution with respect to the disk centre, while that on the right gives the convective exchange of vapour between the disk and the ambient air.

The solution of the above initial-boundary value problem was constructed with the help of the variable separation method, and its final form is

$$\theta^{(I)}(r, t) = \theta_0 - 2RB^{(I)}C \left\{ \frac{t}{t^{(I)}} + \sum_{n=0}^{\infty} \left[1 - \frac{J_0(\lambda_n r/R)}{\lambda_n^2 J_0(\lambda_n)} \right] e^{-\lambda_n^2 \frac{t}{t^{(I)}}} \right\}. \tag{12}$$

In this solution $B^{(I)} = k^{(I)} R / \Lambda^{(I)}$, $C = (Y_n - Y_a)/R$, $t^{(I)} = R^2/D^{(I)}$, and λ_n is the n-th eigenvalue calculated from the characteristic equation

$$J_1(\lambda_n) = 0 \quad \rightarrow \quad \{\lambda_n\} = \{\lambda_1, \lambda_2, \ldots, \}, \tag{13}$$

where J_0 and J_1 are Bessel functions of the first kind of zero and first order, respectively.

The characteristic Eq. 13 and the integral constants were determined by making use of the boundary and initial conditions. For this calculus the orthogonal condition for Bessel functions is applied (see e.g. Moon and Spencer, 1966).

Solution (12) holds for the constant drying rate period ranging in time from the beginning of drying to the critical moisture content, that is, in the time range $0 \le t \le t_s$, where t_s denotes the time, at which the FSP at the disk surface is reached.

3.2 Second (II) period of drying

The moisture content distribution at the end of the constant drying rate period (I) constitutes the initial condition for the falling drying rate period (II).

The distribution of moisture content during the second period is described by the following differential equation and boundary and initial conditions

$$\frac{\partial \theta^{(II)}}{\partial t} = D^{(II)} \left(\frac{\partial^2 \theta^{(II)}}{\partial r^2} + \frac{1}{r} \frac{\partial \theta^{(II)}}{\partial r} \right), \tag{14}$$

$$\frac{\partial \theta^{(II)}}{\partial r}\bigg|_{r=0} = 0, \qquad -\Lambda^{(II)} \frac{\partial \theta^{(II)}}{\partial r}\bigg|_{r=R} = k^{(II)} \left(\theta^{(II)}\big|_{r=R} - \theta_e\right), \tag{15}$$

$$\theta^{(II)}(r, t_s) = \theta^{(I)}(r, t_s), \quad \theta^{(II)}(R, t_s) = \theta^{(I)}(R, t_s) = \theta_s, \tag{16}$$

where θ_e denotes the equilibrium (final) moisture content and θ_s is the moisture content at FSP at the disk surface.

The moisture distribution in the disk during the second period of drying, which is the solution of the above defined initial-boundary value problem, reads

$$\theta^{(II)}(r, t) = \theta_e + 2 \sum_{n=0}^{\infty} \left\{ (b - \theta_e) B^{(II)} - a \left[\left(1 - \frac{4}{\alpha_n^2}\right) B^{(II)} + 2 \right] \right\}$$
$$\times \frac{J_0(\alpha_n r/R)}{[\alpha_n^2 + (B^{(II)})^2] J_0(\alpha_n)} \exp\left(-\alpha_n^2 \frac{t - t_s}{t^{(II)}}\right), \tag{17}$$

where

$$a = \theta^{(I)}(0, t_s) - \theta^{(I)}(R, t_s) \qquad b = \theta^{(I)}(0, t_s),$$

and $B^{(II)} = k^{(II)} R / \Lambda^{(II)}$, $t^{(II)} = R^2 / D^{(II)}$, and α_n is the n-th eigenvalue calculated from the characteristic equation of the form

$$J_1(\alpha_n) = \frac{B^{(II)}}{\alpha_n} J_0(\alpha_n) \quad \rightarrow \quad \{\alpha_n\} = \{\alpha_1, \alpha_2, \ldots\}. \tag{18}$$

Figure 3 presents the distribution of moisture content in the disk in the first (I) and second (II) period of drying.

This figure shows that the moisture content at the disk surface drops very quickly in the first period (I) of drying. The inclination of the curves at the boundary surface is constant, which is characteristic of the constant drying rate period. In spite of the non-uniformity in moisture distribution, no stresses occur if the moisture content is over the FSP. Otherwise, further drying and dropping of the moisture content below the FSP (first at the disk surface) involves shrinkage and generation of stresses.

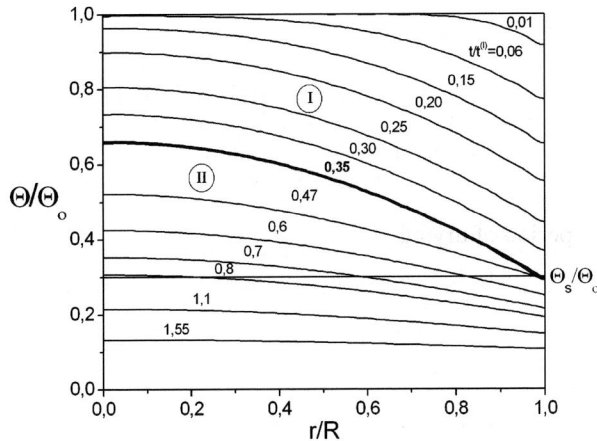

Fig. 3 Distribution of moisture content in the pine-wood disk at several instants of time

The bold line in fig. 3 separates first (I) and second (II) drying periods. It can be seen that the inclination of the curves in the second period becomes smaller and smaller, which means decreasing the drying rate in the course of drying. At the final stage of drying the moisture content becomes uniform in the whole area of the disk and equal to the equilibrium value with the ambient medium, so that the stresses tend to zero. This tendency is shown in Figs. 5 and 6, which present the stress distribution for elastic and, in particular, for viscoelastic material.

Figure 3 enables estimation of radius r_s describing the saturated region (Fig. 2) as a function of time. To this end, we need to read the points of intersection of the curves $\theta(r,t)/\theta_0$ with the line $\theta_s/\theta_0 = $ const for different instants of time. Applying the concept of an Arnold diffusion cell (see Welty, et al. 1976) to a single capillary, we can find a simple relation between drying time and the radius r_s, namely

$$\frac{t - t_s}{t^{(\mathrm{II})}} = \Omega \left[1 - \left(\frac{r_s}{R}\right)^2 \left(1 - \omega \ln \frac{r_s}{R}\right) \right], \qquad (19)$$

where Ω and ω can be estimated on the basis of fig. 3. Choosing two arbitrary values for t and reading appropriate values for r_s from the curves in fig. 3, we obtain the system of two equations, from which we find $\Omega \approx 0.4$ and $\omega \approx 1.05$.

4 Distribution of stresses

The physical relations between stresses and strains for the saturated wood differ from that of unsaturated wood according to the shrinkage term and material coefficients, viz., wood over the FSP does not shrink and has the lowest mechanical strength. Unsaturated wood, on the other hand (moisture content below the FSP), swells or shrinks when the moisture content increases or decreases. The mechanical strength and the mechanical coefficients of wood in this state depend on the moisture content.

Formulae (20) present the physical relations in the saturated region of the cylindrical disk (see Fig. 2)

$$\begin{aligned}
\sigma_{rr} &= 2M^{(S)}\varepsilon_{rr} + A^{(S)}\varepsilon, \\
\sigma_{\varphi\varphi} &= 2M^{(S)}\varepsilon_{\varphi\varphi} + A^{(S)}, \varepsilon \\
\sigma_{zz} &= 0 + A^{(S)}\varepsilon.
\end{aligned} \qquad (20)$$

Physical relation (20) do not contain a shrinkage term and the material coefficients $M^{(S)}$ and $A^{(S)}$ are of the lowest value and constant, i.e., they are independent of the moisture content if it is over the FSP. Thus, relations (20) can be considered as linear ones.

Physical relations in the unsaturated region contain the shrinkage term and the material coefficients are a function of moisture content. In our considerations, however, we use average values for material coefficients, taken as their maximum value when moisture content is minimum and minimum value when the moisture content is close to the FSP. The physical relations for elastic unsaturated wood read

$$\begin{aligned}
\sigma_{rr} &= 2M^{(\mathrm{US})}\varepsilon_{rr} + A^{(\mathrm{US})}\varepsilon - 3K^{(\mathrm{US})}\kappa^{(X)}\theta^{(\mathrm{II})}, \\
\sigma_{\varphi\varphi} &= 2M^{(\mathrm{US})}\varepsilon_{\varphi\varphi} + A^{(\mathrm{US})}\varepsilon - 3K^{(\mathrm{US})}\kappa^{(X)}\theta^{(\mathrm{II})}, \\
\sigma_{zz} &= 0 + A^{(\mathrm{US})}\varepsilon - 3K^{(\mathrm{US})}\kappa^{(X)}\theta^{(\mathrm{II})}.
\end{aligned} \qquad (21)$$

The geometrical relations for the cylindrical disk take the form

$$\varepsilon_{rr} = \frac{\partial u}{\partial r}, \quad \varepsilon_{\varphi\varphi} = \frac{u}{r}, \quad \varepsilon_{zz} = 0, \quad \varepsilon = \varepsilon_{rr} + \varepsilon_{\varphi\varphi} + \varepsilon_{zz}, \tag{22}$$

where $u \equiv u_r$ is the displacement of woody disk in the radial direction.
The balance of internal forces in the cylindrical disk is expressed as follows

$$\frac{\partial \sigma_{rr}}{\partial r} + \frac{\sigma_{rr} - \sigma_{\varphi\varphi}}{r} = 0. \tag{23}$$

Substituting the physical and geometrical relations into the balance of forces (23), we obtain the differential equation for determining displacements and stresses. The integrating constants are determined by making use of the following boundary and compatibility conditions

$$u|_{r=0} = 0, \quad u|_{r=r_S^-} = u|_{r=r_S^+}, \quad \sigma_{rr}|_{r=r_S^-} = \sigma_{rr}|_{r=r_S^+}, \quad \sigma_{rr}|_{r=R} = 0 \tag{24}$$

The solution of the initial-boundary value problem yields the following formulae for the displacement and the stresses in a wood disk dried convectively

– in the saturated region ($0 \leq r \leq r_S$),

$$\frac{u(r,t)}{u_0} = \frac{rR\beta}{(\alpha+\beta)R^2 - (1-\alpha)\beta r_S^2} A(t), \tag{25}$$

$$\frac{\sigma_{rr}(r,t)}{\sigma_0} = \frac{\sigma_{\varphi\varphi}(r,t)}{\sigma_0} = \frac{\alpha(1+\beta)}{(\alpha+\beta)R^2 - (1-\alpha)\beta r_S^2} R^2 A(t), \tag{26}$$

– in the unsaturated region ($r_S \leq r \leq R$),

$$\frac{u(r,t)}{u_0} = \frac{r}{R(1+\beta)} \left[\frac{[(\alpha+\beta)r^2 + (1-\alpha)r_S^2]\beta}{(\alpha+\beta)R^2 - (1-\alpha)\beta r_S^2} \frac{R^2}{r^2} A(t) + B(r,t) \right], \tag{27}$$

$$\frac{\sigma_{rr}(r,t)}{\sigma_0} = \frac{(\alpha+\beta)r^2 - (1-\alpha)\beta r_S^2}{(\alpha+\beta)R^2 - (1-\alpha)\beta r_S^2} \frac{R^2}{r^2} A(t) - B(r,t), \tag{28}$$

$$\frac{\sigma_{\varphi\varphi}(r,t)}{\sigma_0} = \frac{(\alpha+\beta)r^2 + (1-\alpha)\beta r_S^2}{(\alpha+\beta)R^2 - (1-\alpha)\beta r_S^2} \frac{R^2}{r^2} A(t) + B(r,t)$$
$$- \left[\theta^{(II)}(r,t) - \frac{r_S}{r} \theta^{(II)}(r_s,t) \right]. \tag{29}$$

The following notation is introduced in the above formulae

$$\sigma_0 = \frac{2M^{(US)}(2M^{(US)} + 3A^{(US)})}{2M^{(US)} + A^{(US)}} \kappa^{(X)}, \quad u_0 = \frac{2M^{(US)} + 3A^{(US)}}{M^{(US)} + A^{(US)}} \kappa^{(X)} R$$

$$\alpha = \frac{M^{(S)} + A^{(S)}}{M^{(US)} + A^{(US)}}, \quad \beta = \frac{M^{(US)}}{M^{(US)} + A^{(US)}}$$

$$A(t) = \frac{1}{R^2} \int_{r_S}^{R} r \theta^{(II)}(r,t) dr, \quad B(r,t) = \frac{1}{r^2} \int_{r_S}^{r} r \theta^{(II)}(r,t) dr$$

Formulae (25)–(29) describe the displacements and stresses determined on the basis of the elastic model of a wood disk. Stresses determined on the basis of the viscoelastic model follow from the convolution formula (9) and have the form

- in the saturated region ($0 \leq r \leq r_S$),

$$\frac{\sigma_{rr}^{(v)}(r,t)}{\sigma_0} = \frac{\sigma_{\varphi\varphi}^{(v)}(r,t)}{\sigma_0} = \frac{\alpha(1+\beta)}{(\alpha+\beta)R^2 - (1-\alpha)\beta r_S^2} R^2 \left[A(t) - A^{(v)}(t) \right] \quad (30)$$

- in the unsaturated region ($r_S \leq r \leq R$)

$$\frac{\sigma_{rr}^{(v)}(r,t)}{\sigma_0} = \frac{(\alpha+\beta)r^2 - (1-\alpha)\beta r_S^2}{(\alpha+\beta)R^2 - (1-\alpha)\beta r_S^2} \frac{R^2}{r^2} \left[A(t) - A^{(v)}(t) \right]$$
$$- \left[B(r,t) - B^{(v)}(r,t) \right], \quad (31)$$

$$\frac{\sigma_{\varphi\varphi}^{(v)}(r,t)}{\sigma_0} = \frac{(\alpha+\beta)r^2 + (1-\alpha)\beta r_S^2}{(\alpha+\beta)R^2 - (1-\alpha)\beta r_S^2} \frac{R^2}{r^2} \left[A(t) - A^{(v)}(t) \right] + [B(r,t)$$
$$- B^{(v)}(r,t)] - \left[\theta^{(II)}(r,t) - \theta^{(v)}(r,t) \right] + \frac{r_S}{r} \left[\theta^{(II)}(r_S,t) - \theta^{(v)}(r_S,t) \right] \quad (32)$$

where

$$A^{(v)}(t) = \frac{1}{\tau} \int_{t_S}^{t} A(\xi) \exp\left(-\frac{t-\xi}{\tau}\right) d\xi, \quad B^{(v)}(r,t) = \frac{1}{\tau} \int_{t_S}^{t} B(r,\xi) \exp\left(-\frac{t-\xi}{\tau}\right) d\xi$$

$$\theta^{(v)}(r,t) = \frac{1}{\tau} \int_{t_S}^{t} \theta^{(II)}(r,\xi) \exp\left(-\frac{t-\xi}{\tau}\right) d\xi.$$

Although r_S itself changes with time, we have neglected this fact in the above integrations, with the justificiation that the evaporation zone in an actual drying process displaces very slow with time.

Figure 4 shows the displacement $u(R,t)$ of the external surface of the cylinder, illustrating the shrinkage of the cylinder during drying.

It is seen that the cylindrical sample does not shrink at the beginning of drying up to the critical point, that is, in the time period $0 \leq t \leq t_S$. It starts to shrink once the moisture content at the surface reaches the FSP.

Figures 5 and 6 present the distribution of radial and circumferential stresses for both elastic and viscoelastic model of wood.

As we see, the radial stresses are negative in the disk. They vary from zero at the surface to some value at $r = r_S$ and become constant in the saturated region ($0 \leq r \leq r_S$).

The circumferential stresses are tensional at the surface and become compressive at some distance from the surface. These stresses are responsible for crack formation in the dried wood.

The stresses are equal to zero in the first period of drying, that is for $0 \leq t/t^I \leq t_S/t^I = 0.35$, and start to grow in the second period, when the moisture content drops below FSP. They reach a maximum in some instant of drying and then start to decrease. This is visible, in particular, for the viscoelastic model. Moreover, the viscoelastic model reveals the phenomenon of stress reverse (see Fig. 6b). This phenomenon is well known in the drying of wood (see e.g., Milota and Qinglin, 1994).

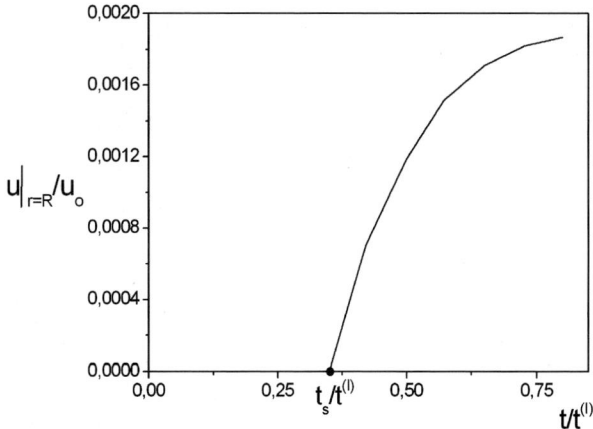

Fig. 4 Shrinkage of the wood disc during drying

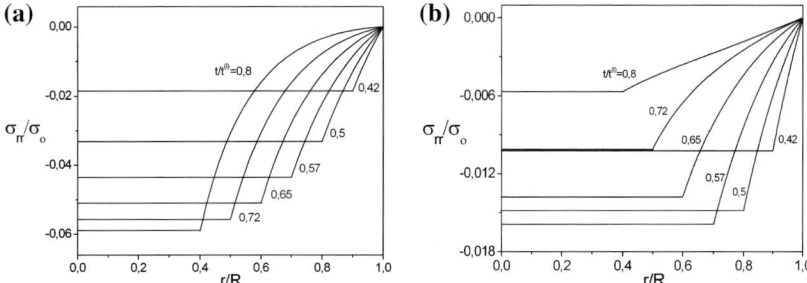

Fig. 5 Radial stresses in woody disk during drying: (**a**) elastic model, (**b**) viscoelastic model

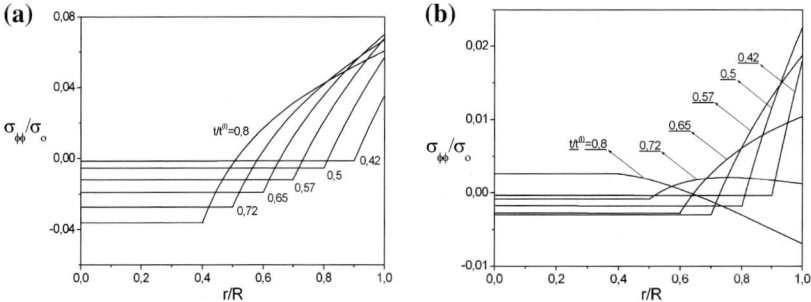

Fig. 6 Circumferential stresses in woody disk: (**a**) elastic model, (**b**) viscoelastic model

It can be explained as follows: when wood dries, the drier surface attempts to shrink but is restrained by the wet core. The surface is stressed in tension and the core in compression. If inelastic strains occur at the surface layer, then, later under the surface with reduced shrinkage, the core dries and attempts to shrink causing the stress state to reverse.

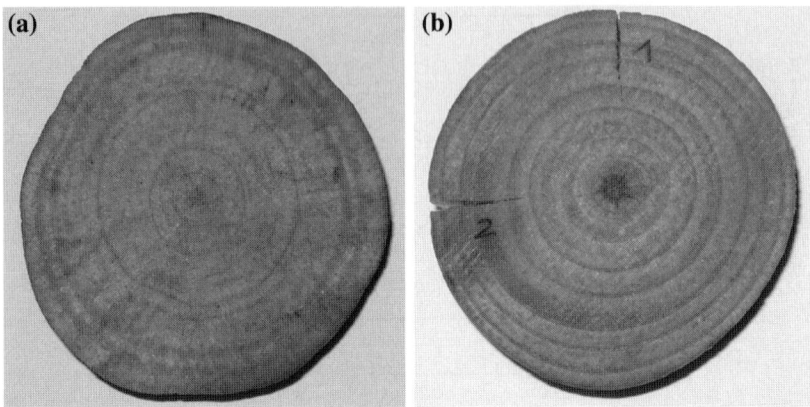

Fig. 7 Photo of wood sample after drying: (**a**) mild drying conditions (80°C), (**b**) harsh drying conditions (100°C)

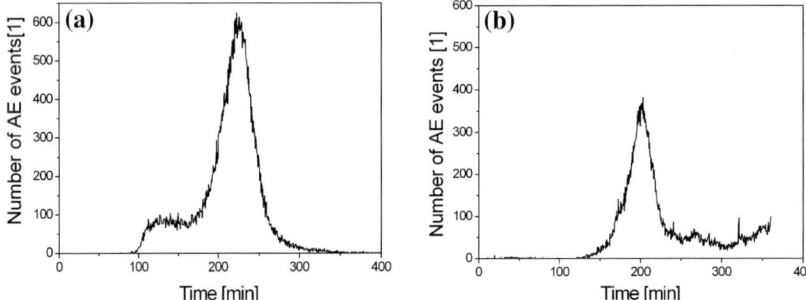

Fig. 8 History of AE event intensities in pine-wood sample dried at: (**a**) 80°C and (**b**) 100°C

5 Acoustic emission by drying of wood

The paper of Kowalski and Smoczkiewicz (2004) presents a way of monitoring of wood destruction with the help of acoustic emission (AE). The studies and results presented there are used here to state whether the acoustic emission reflects the stress development in the dried disk sample estimated theoretically. Figure 7a is a photo of a pine-wood sample after drying in mild drying conditions and Fig. 7b shows one dried in harsh drying conditions.

The samples had cylindrical form: ca. 5 cm in diameter and 1 cm in height. It is seen that the quality of the sample after drying in mild drying conditions is better than the other. Four descriptors were recorded during monitoring of the drying process with the help of the AE: the number of AE events per 30 s time intervals, total number of AE events, the energy of AE events per 30 s time intervals and total energy of AE events.

Figure 8 presents the history of AE event intensities, i.e. the number of acoustic signals per 30 s time intervals for the samples dried at temperatures of 80°C and 100°C

The curves of AE intensity illustrate the moment of drying at which the number of acoustic signals per 30 s time intervals start to increase very rapidly, tending to a maximum. After approaching the maximum, the number of AE signals starts to decrease,

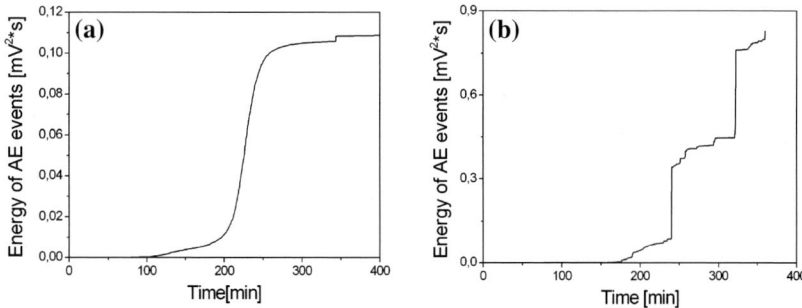

Fig. 9 Total energy emitted by the pine-wood sample dried at: (**a**) 80°C and (**b**) 100°C

which means that the stresses in the dried disk become smaller. Such a conclusion results from analysis of stress distribution in several instants of time (Figs. 5 and 6). In order to compare the time of occurrence of maximum stresses with maximum of the number of AE signals per 30 s time intervals, we have to take the reference time to be $t^{(I)} \approx 2.4 \cdot 10^4$ s.

The numbers of acoustic signals per 30 s time interval well reflect the moment when the stresses start to increase intensely, but they do not reflect the magnitude of the stresses. Note that the curves in Fig. 8 start to increase steeply almost in the same time but the maximum of AE events in drying at temperature 80°C is greater than that in 100°C. This can be explained by the fact that in the latter case the crack developed at some moment (see Fig. 7b), which caused the elastic energy to release and the stresses to decrease for a moment. The crack formation is better reflected by the AE energetic descriptors.

Figure 9 presents the total energy emitted by the pine-wood disk dried at 80C and 100°C.

The descriptor of total AE energy is a very informative one as far as it relates to destruction (cracks) in dried wood. This descriptor from drying at 80°C (Fig. 9a) increases steeply but smoothly from the moment when the wood starts to shrink. No cracks were observed on the surface of the dry disk (Fig. 7a). A quite different plot of total energy was recorded for drying at 100°C. The jump in the increase of total energy at some instants of drying is a response to macroscopic cracks in the wood structure. We can conclude that these two jumps on the curve of total AE energy in Fig. 9b correspond to the two cracks visible in Fig. 7b.

Figure 10 shows the time evolution of circumferential stresses evaluated theoretically and compared with the history of acoustic emission descriptor energy of AE events per 30 s time intervals.

As we see, the enhanced emission of energy of AE events per 30 s time intervals takes place at those moments when the stresses approach their maximum. We can conclude, then, that stresses are responsible for micro- and macrocracks, which are the sources of acoustic emission in dried wood.

6 Final remarks and conclusions

The main aim of this paper was to find a correlation between the stress development determined theoretically in dried pine-wood and the acoustic emission monitored *on line* during drying of such a body. The paper presents the system of equations that

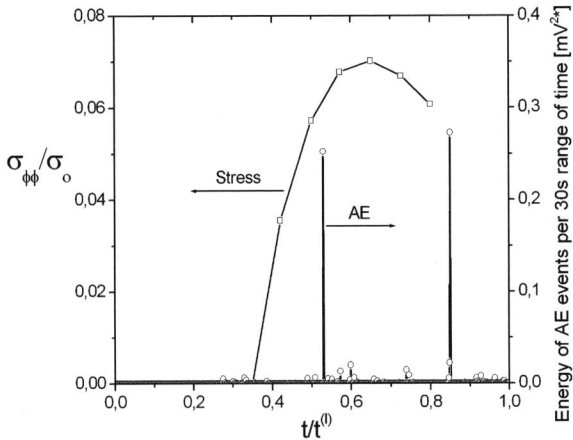

Fig. 10 Comparison of circumferential stress evolution (theory) and the energy of AE events per 30 s time intervals (experiment) during convective drying of disk at 100°C

can be used to calculate stress distribution in a dried body and their evolution in time. The considerations were applied to a cylindrical disk made of pine-wood dried convectively. Such a sample shape used for theoretical analysis was dictated by the experimental testing conditions for AE (Fig. 1b).

Based on the results obtained in this study, we can state that the AE well reflects the development of the drying induced stresses. The descriptor of number of AE events per 30 s time intervals well reflects the start of intense stress generation, but not crack formation. The energetic descriptors, on the other hand, well illustrate the moments of crack occurrence. However, the AE does not allow estimation of the value of the stresses since it only indicates their enhanced development in a given instant of time.

Acknowledgements This work was carried out as a part of the research project No. 3 09TC 037 29 sponsored by Ministry of Science and Education

References

Glijer, L., Matejak, M., Osipiuk, J.: Theory and Technology of Wood Drying, p. 238. PWN, Warszawa (1984)
Kowalski, S.J.: Thermomechanics of Drying Processes, p. 365. Springer Verlag, Heilderberg-Berlin (2003)
Kowalski, S.J., Moliński, W., Musielak, G.: Identification of fracture in dried wood based on theoretical modelling and acoustic emission. Wood Sci. Technol. **38**, 35–52 (2004)
Kowalski, S.J., Rajewska, K.: Drying-induced stresses in elastic and viscoelastic saturated materials. Chem. Eng. Sci. **57**, 3883–3892 (2002)
Kowalski, S.J., Smoczkiewicz, A.: Identification of wood destruction during drying. Maderas: CIENCIA Y TECNOLOGIA **6**(2), 133–144 (2004)
Milota, M.R., Qinglin, W.: Resolution of the stress and strain components during drying of soft wood. In: Proceedings of the 9th International Drying Symposium (IDS 1994), pp. 735–742. Gold Coast, Australia (1994)
Moon, P., Spencer, D.E.: Field Theory for Engineers. p. 600. PWN, Warszawa (in Polish) (1966)
Pang, S.: Modelling of stress development during drying and relief during steaming in *Pinus radiata* lumber. Drying Technol. **18**(8), 1677–1696 (2000)
Welty, J.R., Wicks, C.E., Wilson, R.E.: Fundamentals of Momentum, Heat and Mass Transfer. Willey, New York (1976)

Kinetics of atmospheric freeze-drying of apple

Jan Stawczyk · Sheng Li ·
Dorota Witrowa-Rajchert · Anna Fabisiak

Received: 5 November 2005 / Accepted: 26 March 2006 /
Published online: 1 September 2006
© Springer Science+Business Media B.V. 2006

Abstract We present investigations of the effect of Atmospheric Freeze-Drying kinetics on the quality (dehydration rate, shrinkage, color, and antioxidant properties) of apple cubes. The experimental data are compared with the result of convective and vacuum freeze-drying processes, and suitable operating parameters are determined. The experiments were carried out in an Internet controlled, fully automated heat-pump assisted drying system.

Keywords Kinetics · Atmospheric freeze-drying · Heat-pump · Quality

1 Introduction

The increasing need to approach a balance between high quality of dried product and low-operating cost has induced researchers to investigate the applicability of new techniques of drying. In general, freeze-drying is considered to be the best method for drying food from the product quality point of view, but, due to deep freezing and the low pressures applied, this process is quite expensive. There is a large class of products for which the application of freeze-drying is not economically justified (Wolff and Gibert 1990).

The research and development of such a drying technique as would combine the advantages of both freeze-drying (high-product quality) and convective-drying (low-process costs) gave us one solution, using cold gas with low-water vapor pressure to cause sublimation of moisture from frozen material at or near atmospheric pressure, which is referred to as Atmospheric Freeze-Drying (AFD) (Heldman and Hohner 1974; King and Clark 1987).

J. Stawczyk (✉) · S. Li
Department of Heat and Mass Transfer Processes, Faculty of Process and Environmental Engineering, Technical University of Lodz, Lodz, Poland
e-mail: stawczyk@wipos.p.lodz.pl

D. Witrowa-Rajchert · A. Fabisiak
Department of Food Engineering and Process Management, Warsaw Agricultural University, 02-776 Warsaw, Poland

The main objectives of the research carried out in the Department of Heat and Mass Transfer, Technical University of Łódź, were to design and build a heat pump-assisted, packed bed AFD closed system and investigate the drying kinetics effect on the quality (rehydration kinetics, shrinkage, color, and antioxidant activity) of apple cubes compared with the result of convective and vacuum freeze-drying processes. Some of the present results are discussed in this paper and suitable operating parameters are determined.

2 Project description and experimental set-up

The experiments were carried out in an Internet controlled, fully automated AFD system. The process investigated is a combination of surface freezing ($\sim -10°C$) and maintaining drying (sublimation) at atmospheric pressure. After reaching certain product moisture content (corresponding to the formation of a rigid product, which reduces its shrinkage), an increase of the process temperature up to several or a dozen degrees centigrade above zero is applied. The process is continued until the desired moisture content is reached in the final product. The application of AFD with the use of a heat pump allows us to reduce the energy consumption of the process, while at the same time preserving the advantages of the freeze-drying method. The heat pump was used to cool humid air to a temperature below the dew point, to remove moisture from circulating air and at the same time to control drying agent parameters (Carrington 1996; Kudra and Mujumdar 2002). The heat pump system consists of a refrigerating compressor, air-cooled condenser, evaporator, liquid tank, and a three-way valve that controls evaporator temperature and the depth of cooling and gas drying. A diagram of the experimental rig is shown in Fig. 1.

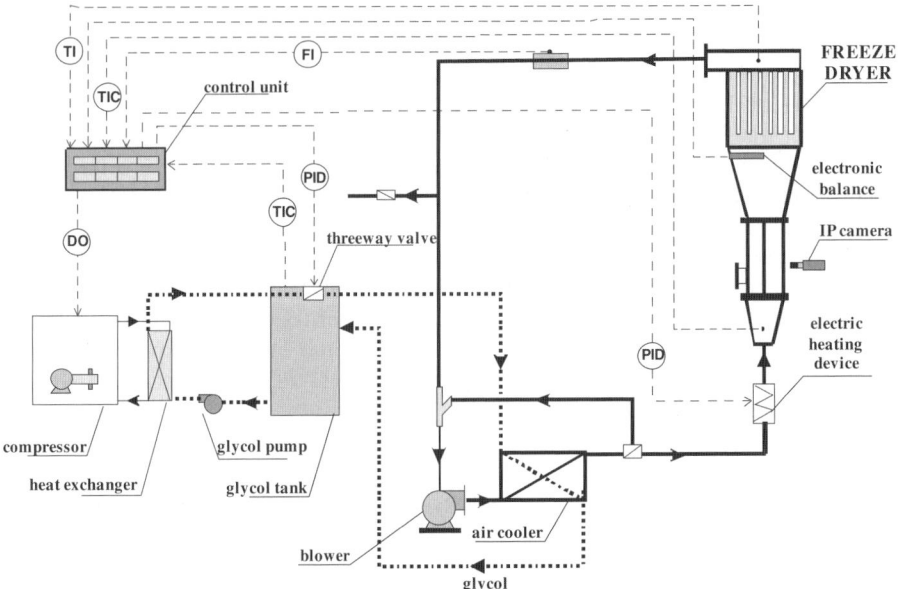

Fig. 1 Diagram of the apparatus for freeze-drying of food products in a closed cycle on a laboratory scale

The experimental rig consists of a cylindrical packed bed dryer with vertical flow of drying gas at a controlled velocity. After leaving the dryer, the gas is directed to the heat pump evaporator cooling system where it is cooled and moisture is removed in the form of mist or frost. The gas dried in this way is directed to the electric heater where it is heated several degrees, thus reducing its relative humidity; it then flows to the drying chamber. This closed cycle eliminates emissions of all types of dusts or odors to the atmosphere.

A data acquisition and control system based on the National Instruments FieldPoint platform was also developed. More information about the control system is available in the paper "Atmospheric Freeze Drying Process Control" (Stawczyk 2005) in the proceedings of this symposium.

3 Atmospheric Freeze-Drying (AFD) kinetics results

It is known that the air inlet temperature plays an important role among AFD operating parameters. Three different temperature (DT) increasing strategies at the same air flow rate (as the process is an internal mass transfer controlled, external hydrodynamics has a relatively small effect, Heldman and Hohner 1974) were implemented, in order to investigate the influence of temperature increase history on drying kinetics. The optimization of the AFD process was adjusted by quality evaluation result from Warsaw Agriculture University.

All 1 cm apple cubes dried in the experiments were from the same kind of fresh apple stored in a refrigerator at 12°C, diced at ambient temperature, and dipped in 300 ml 3% citric acid solution for about 30 min in advance to prevent enzymatic browning during the experiment.

3.1 Inlet air constant temperature option

As a fundamental reference basis, drying kinetics (Fig. 2) for the AFD process in constant temperature (CT) condition was acquired from apple cubes. The drying rate curves show linear characteristics in the run at $-16, -12, -8, -4$, and no constant drying rate period was found (for the sake of simplification the initial surface freezing part was omitted). The drying process was controlled by water vapor diffusion within partially dried porous medium according to the Uniformly Retreating Ice Front (URIF) model. The mechanism of this temperature increase caused a drying rate change mainly connected with the equilibrium of the ice front. The close dependence of drying rate on air temperature was caused by higher ice core temperature, which in turn caused a higher saturated vapor pressure at the ice–vapor interface. The higher vapor pressure at the interface represented an increase in the mass transfer potential and caused a minor increase in the vapor diffusivity due to increasing temperature (Heldman and Hohner 1974).

To calculate the drying rate, here we assume that the value of total raw material surface area was constant, especially below 0°C. This was confirmed by on-line images taken by a network IP camera; major shrinkage increase occurred at the end of the drying process or when the inlet temperature was elevated above 0°C.

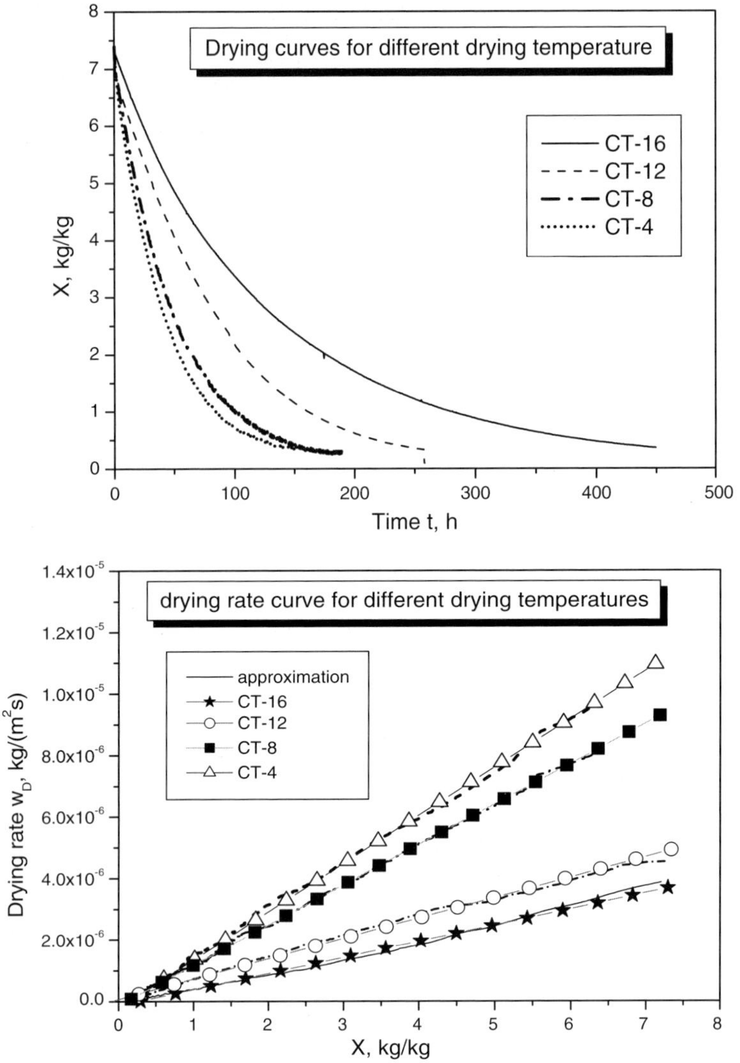

Fig. 2 Drying kinetics of CT option (below 0°C)

3.2 Inlet air different temperature option

A comparison of drying kinetics of apple cubes for CT22 and DT-8/22 options is shown in Fig. 3. Four different drying processes were observed: a prechilling mixed with a thin layer of free water (or ice) diffusion caused by higher initial material temperature (at ambient temperature); an internal mass diffusion-controlled sublimation process; a melting period probably combined with water diffusion among partially dried porous tissue; and a high-temperature (above 0°C) internal moisture transport process controlled by a heat pump. After a temperature jump at $x = 4.76$, the drying rate accelerated from 0.5E-5 to 5E-5 kg/(m²s).

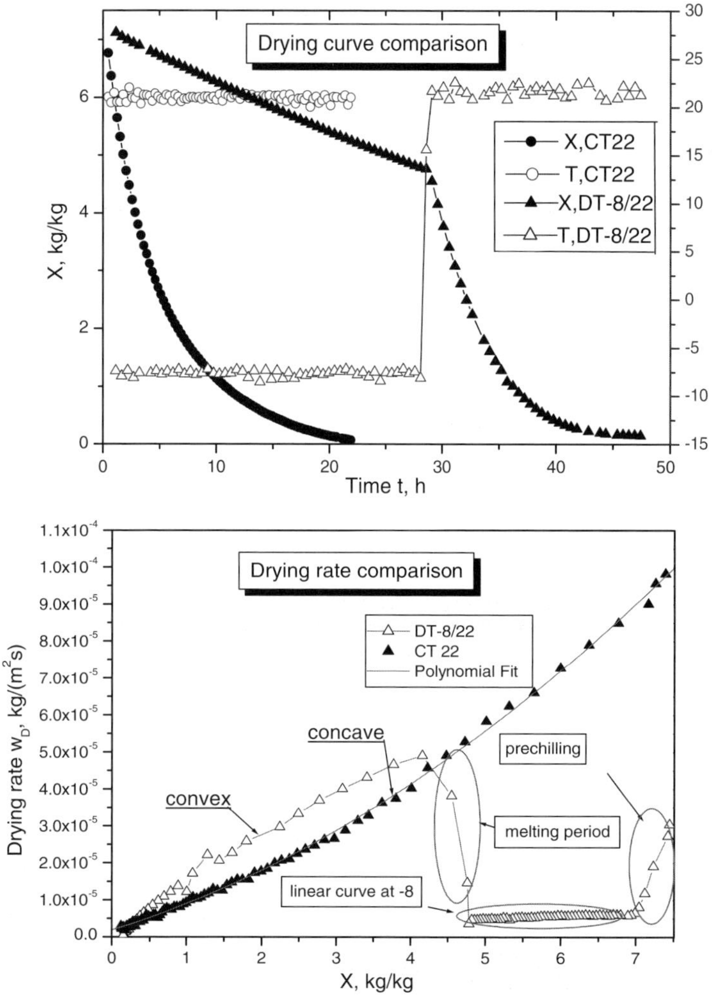

Fig. 3 Comparison of drying kinetics between CT22 and DT -8/22

The drying rate curve DT-8/22 at higher temperature shows a convex shape [however, this was also observed in ascending temperature (AT) drying rate curves], compared with the same part of the CT22 drying curve (concave shape). The convex drying rate curve presents more the characteristics of thin leather and textiles rather than the CT22 concave characteristic, which are those of capillary-porous bodies with small specific evaporation surfaces (ceramics, clay) (Strumillo and Kudra 1986). This kind of difference indicates that the surface freezing process (below 0°C) preserved the advantages of freeze-drying (small shrinkage), and gave the semi-dried products higher porosity than those of higher temperature convective-drying process (above 0°C) at the same moisture content.

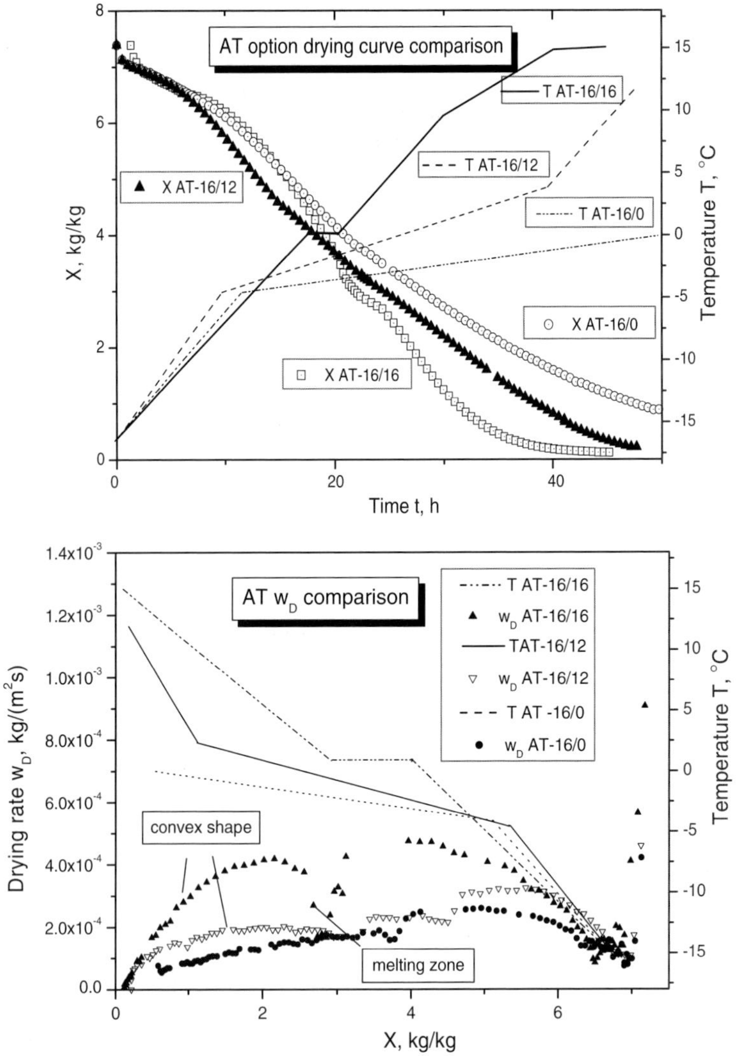

Fig. 4 (**a**) Comparison of drying kinetics for AT option. (**b**) Comparison of drying kinetics for AT-16/-4/4/22 and AT-16/-4/4/22quick

3.3 Inlet air ascending temperature option

In order to avoid potential thermal damage caused by an inlet temperature jump (DT), and based on several trial results, an ascending temperature (AT) option (shown in Fig. 4a) was tried. The difference among these three runs was the rate of temperature increase around 0°C. An ascending inlet temperature was tried to maintain a stable drying rate during the whole drying process to obtain an economic AFD process and mostly for better quality. In our work, more experiments were done on apple to develop a suitable T-inlet curve.

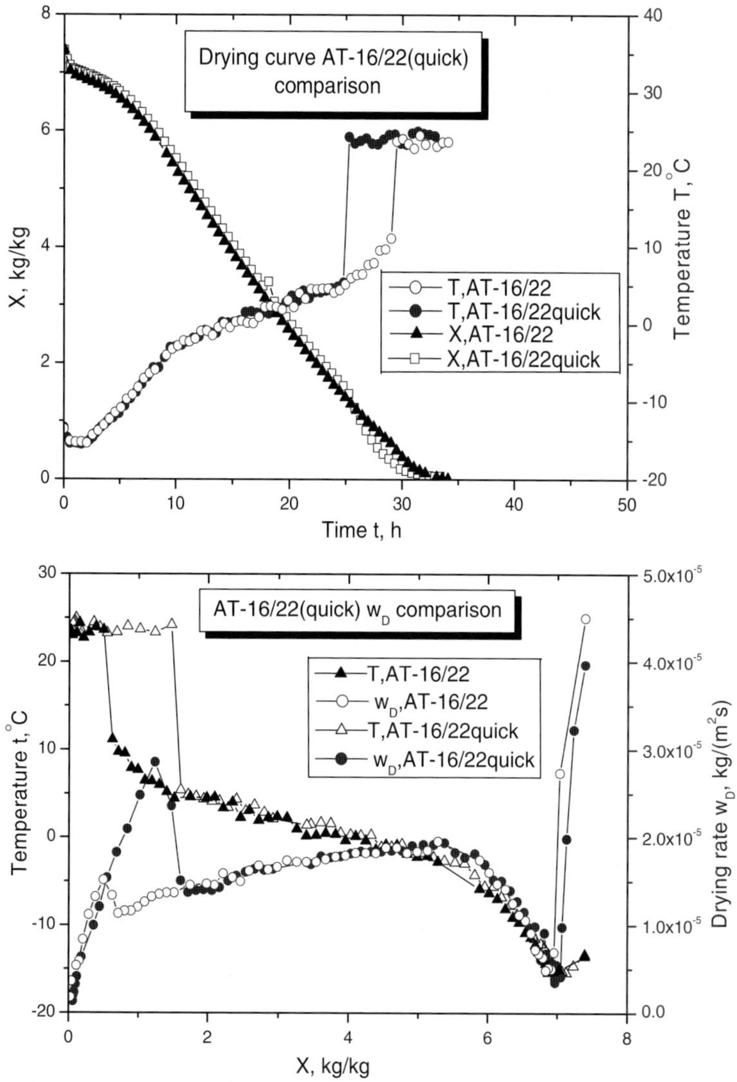

Fig. 4 continued

The melting period at high-temperature increase was detected. This occurs because melting consumed most of the exchanged heat and only a small fraction was consumed by moisture evaporation. This can be seen as a drop of the drying curve between two peaks. At this period, some deterioration of the tissue structure (collapse, shrinkage break and soluble ingredients redistribution, etc.) might occur. It is advised to increase temperature carefully when the drying process reaches the melting point. An economically adjusted AT-16/-4/4/22 strategy (with a middle melting region and higher inlet temperature at the end of the process compared with AT-16/16 and AT-16/12) yielded a better quality of final dried apple cubes than those from the CT, DT, and AT runs (shown in Fig. 4b).

4 Quality evaluation

The quality evaluation results were used as a key concern to readjust and optimize operating parameters. Below, we present the analysis of results in terms of rehydration kinetics, shrinkage, hygroscopic characteristics, color, and antioxidant activity.

4.1 Rehydration kinetics

Rehydration was carried out by flooding a known mass of dry material with distilled water at the temperature of 20°C. After 0.5, 1, 2, 3, 4, and 5 h, the sample was separated from water on a sieve, blotted with filter paper, and weighed. Dry matter content was determined according to the PN-90/A-75101 standard. Relative mass increase in the tissue (Fig. 5) was analyzed.

Initially, the high porosity of the freeze-dried product causes an intense mass increase induced by capillary absorption and filling of the porous material with water. During further rehydration, the mass increase is insignificant, growing about fourfold after 5 h of the process. Similar water absorption is characteristic of dried products obtained using CT-16 and CT-12. The dried product obtained in all other experiments absorbs slightly less water than the freeze-dried product at the beginning of the process, and the two rehydration kinetics are similar to that of product dried by convective-drying at 70°C. Mass increase after 5 h rehydration is on a similar level for almost all products tested. Deviations may be attributed to the differences in material taken for experiments.

4.2 Relative dry matter content

The smallest loss of soluble components in the dry matter was observed in a freeze-dried product, in which ca. 40% of the initial dry matter remained after 5 h rehydration. This provides evidence of the smallest failure in the permeability of cell membranes induced by water removal (shown in Fig. 6). More soluble components (65–80%) diffused from all other dried materials; some differences in this loss were observed depending on drying parameters and techniques: DT-4/22, AT-16/-4/4/22, and (quick) AFD runs are better than the convection result, whilst AT-16/16 has the lowest dry matter content. When the higher moisture contents of dry product at lower temperature were taken into account (for CT-8, $X = 0.43$, CT-4, $X = 0.28$ higher than the moisture content from Sublimation 30 and Convection 70), the relative-dry matter content for the AFD result in some low CT conditions will also be higher. This might be a potential reason for the "deviations" in the density and hygroscopic kinetics.

4.3 Hygroscopic properties

It follows from the analysis of water vapor sorption kinetics that all products dried at low temperatures have better hygroscopic properties than the product dried by convection, and worse than the freeze-dried product (Fig. 7). Dried product (CT-16) has vapor absorption properties closest to the freeze-dried product, as is also confirmed by its density measurement. Dried products of CT-4, CT-8, and CT-12 have similar hygroscopic properties to the product dried by convection (as confirmed by density), like dried product DT-8/22 (however, this is not confirmed by density measurements).

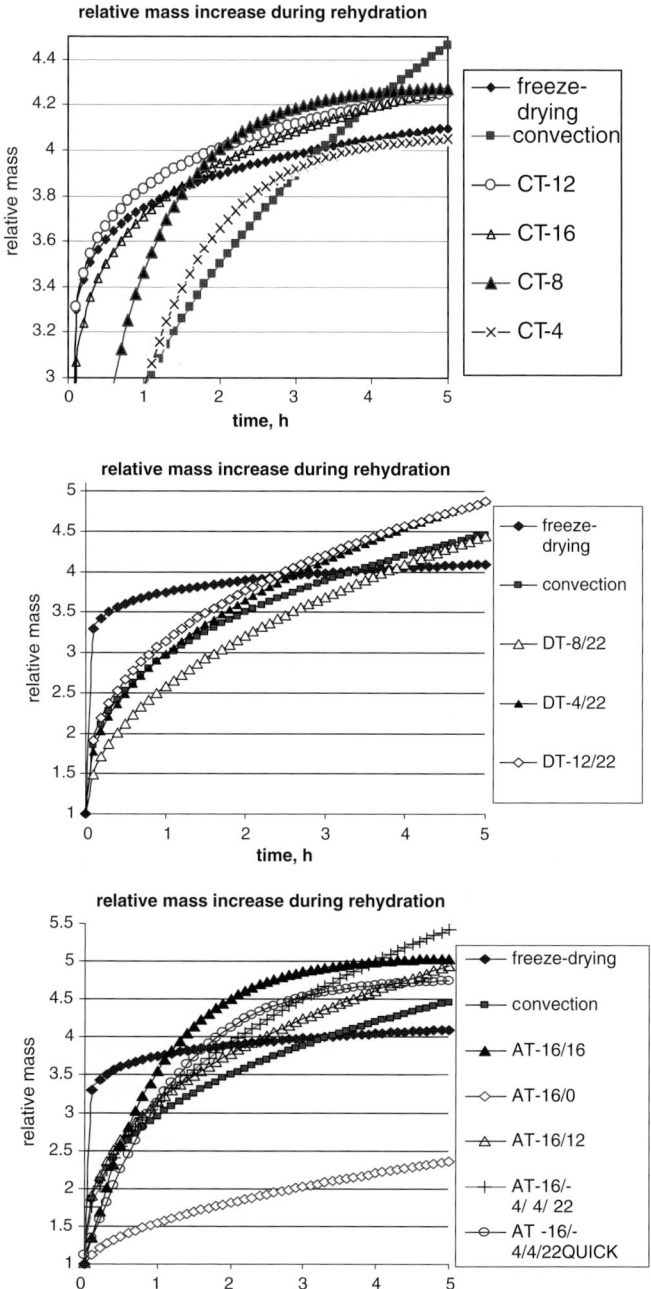

Fig. 5 Relative mass increase during rehydration

Fig. 6 Relative mass increase and dry matter content during rehydration

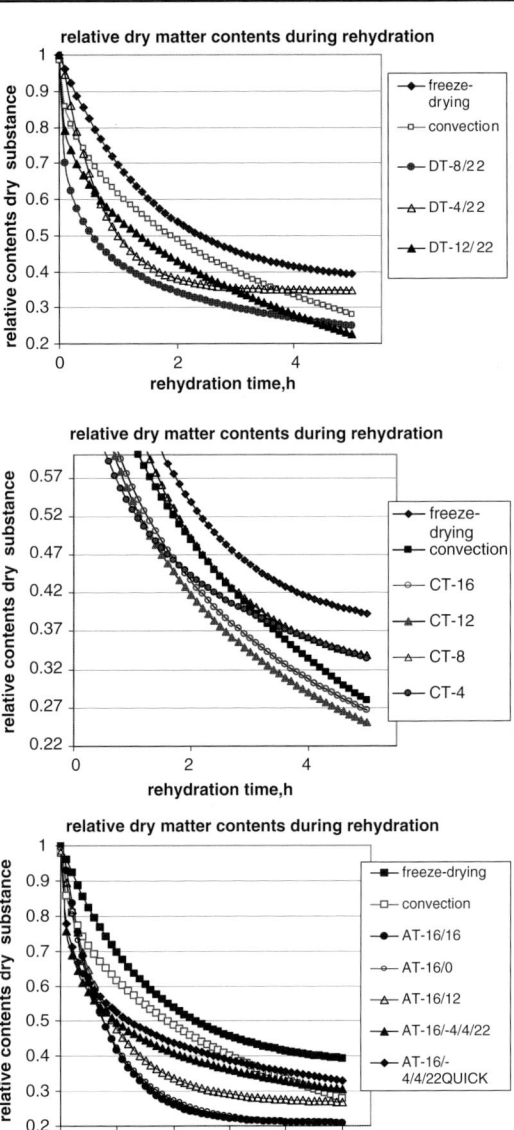

The high-water absorption of the three other dried products from AT-16/-4/4/22, DT-4/22, and DT-12/22 is difficult to explain. The fact that they were not dried up probably means that the inner structure and compounds included in the tissue were damaged to a lesser extent than in other products.

4.4 Shrinkage, density, and antioxidant properties

In the case of low-temperature dried products, shrinkage (percentage of decreased volume of dried product to the raw volume before drying) was calculated under the

Fig. 7 Sorption kinetics of water vapor

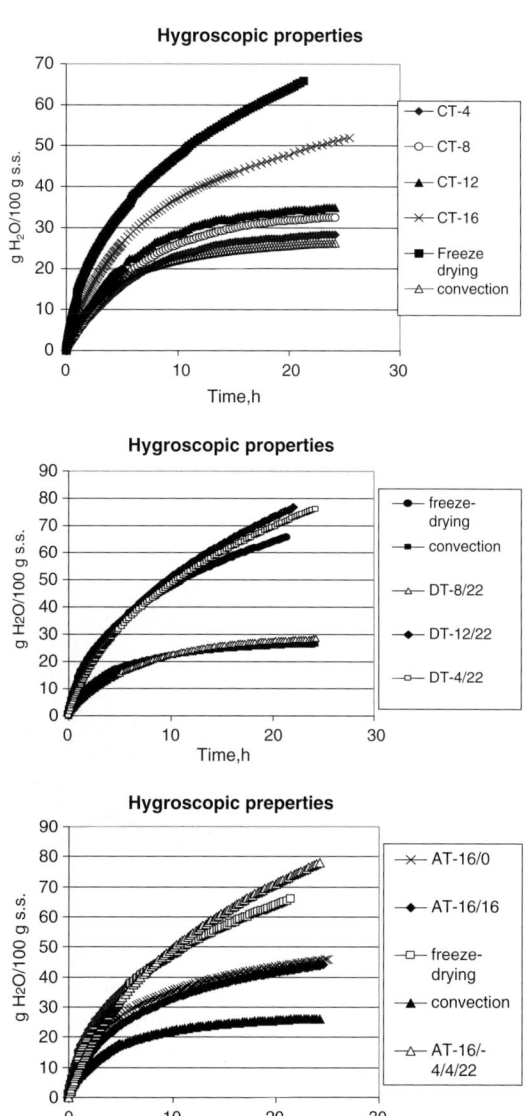

assumption that cube volume prior to drying was 1 cm³, which might not be true in fact. The analysis of this parameter is thus done on the basis of dried product density. The results are shown in Table 1.The lower it is, the smaller is the raw material shrinkage and the more porous the product—the freeze-dried product density is only ca. $0.17 \, g/cm^3$. Drying by convection leads to stiffening of cell walls, which causes formation of a stiff external layer, thus increasing the density (up to $0.45 \, g/cm^3$) and shrinkage and decreasing the material porosity.

Among the low-temperature dried products, the smallest density was shown by the products dried from CT-16, DT-8/22, and AT-16/-4/4/22. Others were characterized by a density similar to that of the convective dried product. The high density of certain

Table 1 Shrinkage, density, and antioxidant properties

Dry product content	Density (g/cm^3)	Shrinkage (%)	Reduction in relation to raw apple (%)	
			Oxidative activity	Polyphenol
Freeze-dried 30°C	0.17 ± 0.02	33.8	3.20	2.40
Convection 70°C	0.45 ± 0.12	73.2	30.4	35.6
CT-4	0.49 ± 0.12	70.9	19.4	12.6
CT-8	0.49 ± 0.03	74.2	20.0	28.6
CT-12	0.54 ± 0.03	72.3	7.8	23.9
CT-16	0.38 ± 0.08	73.3	21.0	5.30
DT-4/22	0.42 ± 0.06	75.6	23.1	32.0
DT-8/22	0.35 ± 0.08	79.5	12.2	46.4
DT-12/22	0.37 ± 0.03	73.6	41.0	39.5
AT-16/0	0.64 ± 0.11	72.3	26.2	4.22
AT-16/12	0.42 ± 0.06	60.7	16.1	19.1
AT-16/16	0.47 ± 0.10	66.7	6.0	3.30
T-16/-4/4/22	0.37 ± 0.04	68.7	7.51	20.6
AT-16/-4/4/22quick	0.40 ± 0.08	72.9	12.1	21.1

Fig. 8 Pictures of fresh and dried apple cubes of CT-16

dried products is directly related to the fact that they were not dried completely (high-equilibrium moisture content at lower temperature). Many researchers have found that porosity increases significantly in the final stages of drying due to structure stiffening at the end of the process and decreasing shrinkage. This is obviously connected with volume decrease. Sample pictures of fresh and dried apple cubes are shown in Fig. 8.

Antioxidant properties of the tested products were determined by a method based on specifying a reduction degree (scavenging) of DPPH radicals by antioxidants. The scavenging efficiency of free radicals by the tested products was given in the form of coefficient IC_{50}, which defines the amount of dry matter required for a 50% reduction of free radicals. Determination of polyphenol compounds, which greatly affect the antioxidant activity in apples, was done by the Folin–Ciocalteau method. The highest efficiency of free radical scavenging characterized the freeze-dried product, for which a statistically significant decrease of antioxidant activity in relation to fresh apple (ca. 3.2%) was found. At the same time, despite a slight decrease, the content of polyphenols in this dry product did not differ significantly from their content in the raw material prior to drying. During the drying by convection, the antioxidant activity of apples decreased remarkably, reaching 70% of raw material activity before drying. Similarly, the content of polyphenols was reduced by about 35%.

Low-temperature drying at CT, DT, and AT also caused a decrease of antioxidant activity and polyphenol content, but it was not so significant as in the case of drying at high temperatures, but it was higher than for the freeze-dried product. The antioxidant activity decreased by about 8–20%, while the content of polyphenols reduced by ca. 12–30%. The other low-temperature dried products had a similar antioxidant activity. The highest antioxidant activity and the greatest content of polyphenols, close to that of the freeze-dried product, were shown by the dried product denoted AT-16/16.

4.5 Color

Freeze-drying does not induce remarkable changes (although they are statistically significant) in the color of dry products and apple brightness is the closest to the fresh material brightness. Similar, although slightly lower brightness values (the difference is statistically insignificant), were obtained for the product CT-16, CT-12, CT-4, and CT-8. Water removal as a result of drying by convection resulted in color deterioration caused by biochemical reactions (enzymatic browning). Dried products CAT-16/16 and CAT-16/0 were characterized by a lower brightness L parameter than the product dried by convection. The other low-temperature products DT-8/22 and AT-16/12 had brightness similar to that of dried products obtained at CTs. It should be mentioned, however, that this measurement for low-temperature dried products is not made immediately after drying, and a storage period surely affects the color of the product.

5 Conclusions

Mass diffusion controls the AFD process of apple dewatering at air temperatures below 0°C.

The AFD process of apple dewatering run at temperatures around −10°C leads to a highly porous product structure. The same process performed at temperatures around 0°C results in deterioration of product quality.

The quality evaluation of apple cubes shows that dried products of AFD at lower temperature have similar characteristic of rehydration kinetics and hygroscopic properties to the product obtained from vacuum freeze-drying. The AFD product results have a statistically higher value of antioxidant activity and polyphenol content compared with convective-drying result.

The optimum drying trajectories for apple cubes were found for the AT drying mode, where a middle melting region and constant drying rate occur.

Acknowledgments The present paper has been prepared within the research project no. 4 T09C 048 23 entitled "Freeze Drying of Food Products in a Closed System" sponsored by the State Committee for Scientific Research in 2002.

References

Carrington, C.G.: An empirical model for a heat pump dehumidifier drier. Int. J. Energy Res. **20**, 8530–869 (1996)

Heldman, D. R., Hohner G.A.: An analysis of atmospheric freeze-drying. J. Food Sci. **39**, 147–155 (1974)

King, C.J., Clark, J.P.: System for freeze-drying. U.S. Patent No. 4,697,358 (1987)

Kudra, T., Mujumdar, A.S.: Advanced Drying Technologies, Marcel Dekker Publications, New York (2002)

Wolff, E., Gibert, H.: Atmospheric freeze-drying, Part 1: design, experimental investigation and energy-saving advantages. Drying Technol. **8**(2), 385–404 (1990)

Stawczyk, J.: Atmospheric freeze-drying process control. In: Proceedings of 11th Polish Drying Symposium, on CDROM, 2005

Strumillo, C., Kudra, T.: Drying: Principles, Applications and Design. Gordon and Breach Science Publications, London (1986)

Microwave drying of various shape particles suspended in an air stream

Michal Araszkiewicz · Antoni Koziol · Anita Lupinska · Michal Lupinski

Received: 30 October 2005 / Accepted: 26 March 2006 /
Published online: 13 December 2006
© Springer Science+Business Media B.V. 2006

Abstract Fluidization is an efficient way to dry granular materials. Incorporating microwave heating into the fluidization makes the overall drying process shorter, and the quality of the final products can be improved. However, in order to understand the mechanisms of water removal, an exact knowledge of changes inside the dried material is necessary. The temperature and moisture distribution pattern within the heated material should be identified and analyzed. Unfortunately, the microwave environment makes the measurements very difficult. This paper gives new information on the temperature distribution inside small particles of various shapes dried with microwaves. The tests were carried out in a laboratory-scale, fluid-bed dryer equipped with a microwave source. Five different shapes were examined: sphere, cylinder, half-cylinder, rectangular prism, and prism with triangle base. All particles tested were suspended in an air stream and heated with microwaves. The internal temperature distribution has been analyzed in each case. The rate of drying is also presented and discussed for every case tested.

Keywords Microwaves · Drying · Fluidization · Granular materials · Temperature distribution · Heat transfer

Nomenclature
E   Electric field strength, V/m
f   Frequency, Hz
P   Power, W
X   Moisture fracture

Greek
ε Porosity
ε'' Microwave energy dissipation coefficient, F/m

M. Araszkiewicz (✉)· A. Koziol · A. Lupinska · M. Lupinski
Division of Chemical and Biochemical Processes, Faculty of Chemistry, Wroclaw University of Technology, ul. Norwida 4/6, 50-373 Wroclaw, Poland
e-mail: michal.araszkiewicz@pwr.wroc.pl

ε_0 Dielectric constant in vacuum, $8,85 \cdot 10^{-12}$ F/m
π Pi number

Subscripts
l Liquid
s Solid

1 Introduction

Microwave-assisted fluid bed drying is an alternative method for the rapid, efficient dehydration of granular material. Since their discovery in the mid-1940s, microwaves have gained increasing attention as an optional and very efficient supplement to the standard heating technologies in many industries. Microwave heating is used, in particular, in the food industry for baking, blanching, cooking, dehydration, drying, pasteurization, and sterilization (Dincov 2004).

Despite almost 60 years of experience with microwaves, there are still many questions and doubts. Moreover, conducting experiments is not especially easy due to the ability of microwaves to exclude most of the common measurement techniques widely used in drying science.

This work presents our attempt to exhibit, describe, and discuss the influence of material shape on the temperature distribution within a solid during fluid bed drying assisted with microwave heating. Additionally, the paper focuses mainly on the experimental results; some aspects of the modeling will also be touched. Of course, the author's main goal was to determine the changes within the material that occur during microwave-assisted drying in a fluid bed environment.

The most important feature of fluid bed drying compared to the standard drying technique is the free movement of the dried/heated material inside the drying cavity, in both horizontal and vertical directions. This fact makes the whole process much more complicated and difficult to analyze. Additionally, the size of the dried material is smaller than the wavelength of the microwaves. The next section very briefly discusses the theoretical background of the microwave-assisted drying process and formulates the aim of the experiments. The experimental set up is described and the measurement technique is presented. The most interesting results are reported and discussed.

2 Volumetric heating

Microwave-assisted drying is also known as volumetric heating, which means that the whole volume of the heated material placed inside the cavity is exposed to the influence of the microwave energy. The temperature rise appears in the whole volume of the heated material, so the risk of overheating the material can be minimized. The quantity of the energy that reaches the core of the heated material no longer depends on the thermal diffusivity and surface temperature (Meredith 1998). The heat generation can be estimated from the well-known equation:

$$P = 2\pi f \varepsilon_0 \varepsilon''_X E_r^2, \tag{1}$$

where the efficiency of the heat generation (i.e. heat power dissipated within the material $P[W]$) is proportional to the square of the electric field intensity (E) at the

exact location (subscript r) of the material and to the loss factor of the material (ε''). Please note the subscript X, which implies the dependence of the loss factor on the material's moisture content.

The whole volume of the porous particle is a mixture of the three phases: solid matrix, liquid water trapped within the void (pores), and air. The resultant loss factor of this system can be obtained from the equation (Wang et al. 2000):

$$\varepsilon''_X = \varepsilon''_l \varepsilon X + \varepsilon''_s (1 - \varepsilon), \qquad (2)$$

where ε is the porosity of the heated material, X denotes the moisture content and subscripts l and s indicate the loss factors of liquid (l) or solid (s), respectively. As a matter of fact, the values of the loss factor of many materials are much smaller than the loss factor of water. Additionally, the loss factor can also change its value at different temperatures and the exact values can sometimes be hard to obtain for some kinds of materials. According to equation (2), the intensity of heat generation will decrease related to the decrement of water during the time of drying.

In order to calculate the P-value properly, a knowledge of the exact value of the electric field intensity is necessary. This parameter is hard to obtain, especially if we consider its change within the exact location in the material analyzed. So, in order to obtain an accurate value of the heat generation, two main variables are necessary: moisture content and exact depth inside the material (even if the material's size is smaller than the microwave wavelength, which is approximately 12 cm for 2450 MHz).

There are some papers (e.g. Datta 1990, Metaxas & Meredith 1983) that describe how the problem of temperature distribution inside the heated material is connected with its shape. The specific case takes place during microwave-assisted drying of a small sphere, which is supposed to act like a lens and focus the electromagnetic waves inside (Schlunder 1993; Araszkiewicz et al. 2004; Remmen et al. 1996). In this particular case, the temperature maximum appears to be at the geometrical centre of the particle. This phenomenon seems to arise independently of the moisture content. This process can lead to the risk of overheating the core of the material, because a highly non-uniform temperature distribution is observed in such cases. Therefore, a knowledge of the changes within the material during microwave-assisted drying seems to be very important.

3 Heat and mass transfer

The most important difference between microwave-assisted and convective drying is the direction of the heat transfer. During convective heating, the heat flux reaches the surface of the dried solid and is transferred via the solid matrix to the deeper parts of the material (Fig. 1). During the drying time, the surface of the material is continuously exposed to the higher temperature and the risk of shrinkage and cracking due to the rapid dehydration of this area is very probable. This dry part of the solid also restrains the water transfer from the material's core. Therefore, the common problem can be the presence of still wet solid areas within the theoretically dry material. Moreover, the second stage of drying takes a long time compared to microwave-assisted drying. The intensity of temperature rise within the material is rather constant during the whole process, mostly independently of the moisture content.

The microwave-assisted drying process differs essentially from that of convection. The source of the heat in this case is not in the surroundings of the heated material,

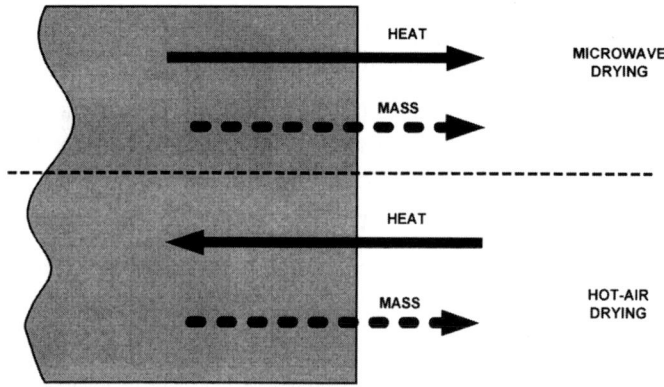

Fig. 1 Direction of the heat and mass transfer during microwave assisted and hot air drying

but within the dried body. The wet material contains the heat source as long as water is still present within the solid matrix. The highest temperature appears in the centre of the heated material and the heat flows towards the outer regions of the material. The air changes its role in the drying process: it is no longer the source of the heat energy, rather, it becomes the cooling factor, especially if we consider a very common situation when the material may have higher temperature than the surrounding air. The heat and mass fluxes align in the same direction, i.e. from inside to outside. The central part of the material dries faster than the surface, where the moisture film is refreshed with the water from the material's interior. In general, the use of microwave energy brings about a meaningful shortening of the drying time compared to traditional heating with hot air (Araszkiewicz et al. 2003). Of course, the intensity of the heat generation is very sensitive to the removal of the moisture content within the material (in most cases). Lack of water within the material during drying reduces the heat generation and its intensity. The solids temperature decreases to the level of the surroundings (air).

This work is a continuation of the author's previous paper, which examined the moisture and temperature distributions within spherical particles of various dimensions (Araszkiewicz et al. 2003). The mathematical model of the heat transfer within a single spherical porous particle gave the correct results under the condition that the electrical field intensity within the solid matrix is non-uniform (Koziol et al. 2005). The previous studies focused on the temperature changes that occurred within the spherical porous particles, although the real particles did not have such regular shapes. This work is an introduction to the studies on the temperature and moisture distributions within small porous particles with various shapes, dried in fluid bed conditions with the application of microwave energy.

4 Objectives of the study

The main objective of this work was to determine the temperature distribution inside porous particles of various shapes. The rate of drying was analyzed in each case. The tested particles had different shapes but approximately the same mass and volume. The dimensions of the tested particles were smaller than the average length of the

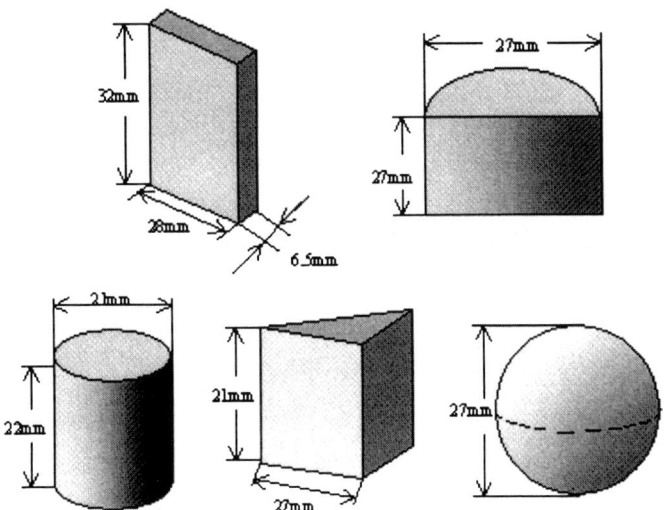

Fig. 2 Tested shapes

wavelength (12 cm at 2450 MHz). Thus, the authors could assume that the whole volume of the material was easily penetrated by microwaves (Meredith 1998). Five different shapes were examined: sphere, cylinder, half-cylinder, rectangular prism and prism with a triangle base (Fig. 2). All experiments presented in this paper were carried out in the same manner and under identical conditions.

5 Experimental set-up

The experimental set-up was the laboratory-scale microwave dryer shown in Fig. 3. The single test particle was freely floated in the air stream and could move in both vertical and horizontal direction within the quartz glass column inserted into the microwave cavity (Fig. 3). The microwave source worked in continuous mode with a specified level of nominal power (from 100 W to 800 W). The nominal microwave power used was 500 W. The air temperature was 40°C. The experiments were conducted intermittently. After a suitable drying time the microwave heating was stopped, the material was taken out of the drying cavity (A in Fig. 4) and measurements of temperature and weight loss were taken. The temperature has been measured in two ways: with a thin thermocouple (K-type, 0.5 mm diameter) inserted at a specific depth inside the tested sphere, and with an infrared photograph of the cross-section of the particle (B, C on the Fig. 4), in order to obtain a general view of the temperature distribution within the particle. Taking the IR photo involved cutting the particle in half along the longest axis of symmetry. The data from the camera were transferred to a PC, where the thermal picture of the material was constructed (D in Fig. 4). The temperature distribution in the cross-section of the sample was then elaborated with software (E and F in Fig. 4). The thermal pictures of the cross-section of the material were taken after a specified heating time in all cases.

The tested particles were made from the same material: gypsum. This sort of material allows the researcher to make a number of identical particles with various shapes.

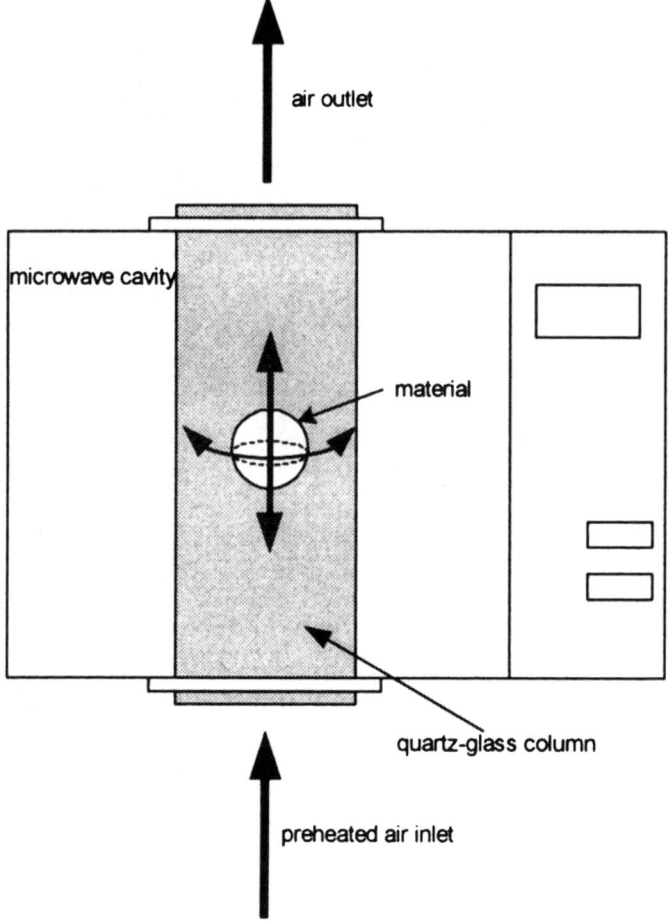

Fig. 3 Experimental set-up

They were porous and their internal structure was uniform. The material was kept in water for a specific time in order to achieve certain initial moisture content, which was almost identical. All particles examined had the same mass, but their surface area changes due to their different shapes.

6 Results

6.1 Moisture removal

The removal of moisture is presented in Fig. 5. Although the moisture removal was similar in all cases examined, the most efficient moisture removal appeared during drying of the rectangular prism during the initial period of the process (0–200 s in Fig. 5). As heating proceeded, the fastest moisture removal occurred in the (triangle) prism case. The spherical particle had the longest drying time.

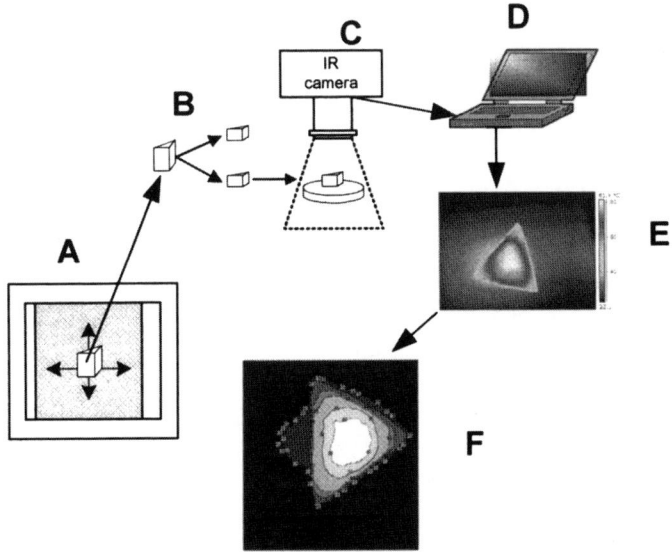

Fig. 4 Measurement technique: internal temperature distribution

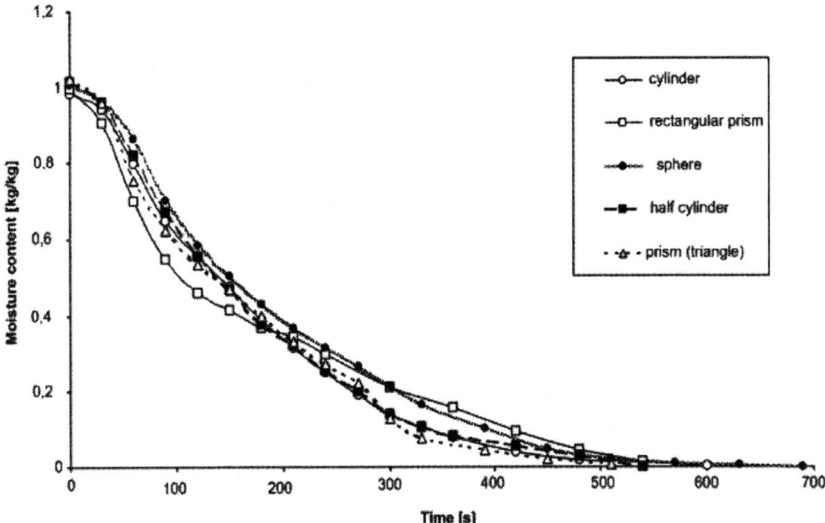

Fig. 5 Moisture content removal versus time

6.2 Rate of drying

The rate of microwave drying is much faster than traditional hot air drying. The most intense water removal takes place at the beginning of the process (Fig. 6), when the heat generation is the most intense. The particles with the greatest surface area were dried at the highest drying rates. This was observed during drying of the rectangular

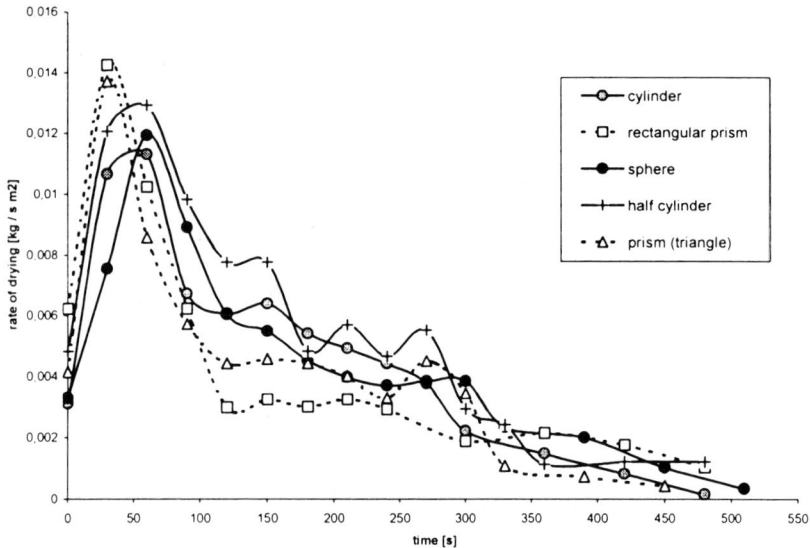

Fig. 6 Rate of drying

prism (Fig. 6), where the drying rate at the beginning of the process is the highest and soon after that decreases almost three-fold. It seems that the particles with greatest surface area (rectangular prism and prism) dry faster and more rapidly than the remainder of the tested particles. The drying rates of the three other particle shapes are smaller and the overall drying process seems to proceed more gently.

6.3 Temperature distribution

During microwave heating, the temperature growth is directly connected with the heat generation, which depends on the moisture content. The highest temperature appears at the beginning of the drying process. The temperature decreases simultaneously with moisture removal. The material temperature reaches the air temperature (40°C) at the end of drying. The shape of the temperature curves is similar in the cases considered. The main differences appear between the core and surface temperatures. This may be connected with the material shape, especially its surface area, which provides better water removal from the particle surface. The largest differences between the core and surface temperatures were noted during drying of the cylinder (Fig. 7), sphere (Fig. 8), and prism (Fig. 9), respectively.

The difference between core and surface temperature in the sphere (Fig. 8) was almost 25°C. The temperature of the centre reaches 90°C during the first 2 min of heating. After that time, the temperature dropped to the temperature of the surroundings. The thermal pictures of the cross-section gave slightly higher temperature readings (Fig. 10). This difference in temperature readings appears in all cases discussed. The reason for these differences is material cooling during thermocouple measurement, which took more time than making an IR image.

The temperature distribution in the cylinder (Fig. 7) is similar to that in the sphere. The differences between the centre and the surface temperatures appear to be even larger than in the sphere case. The temperature in the cylinder core reached almost

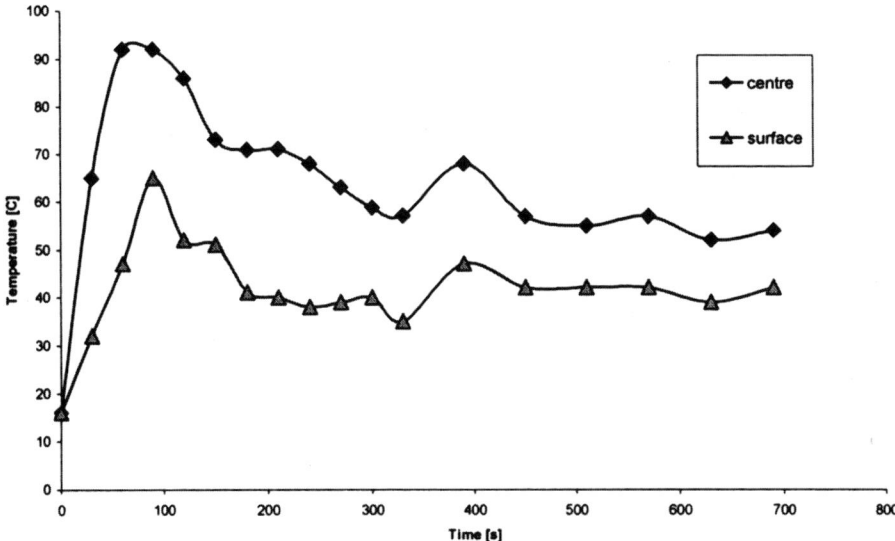

Fig. 7 Temperature distribution in cylindrical particle

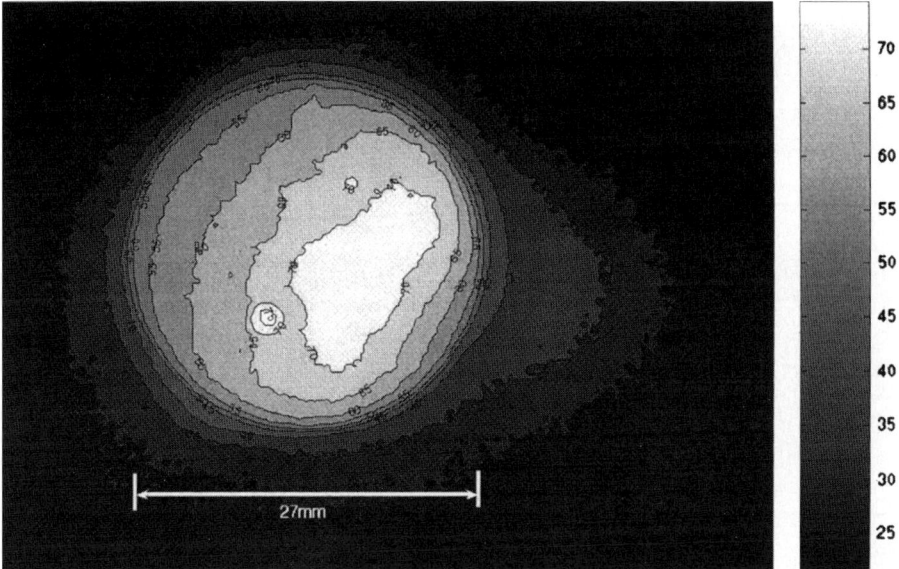

Fig. 8 Temperature distribution in spherical particle

100°C in the first 100 s of drying. There were no differences between the base surface and the cylindrical face surface temperatures. The core temperature remained at the 70°C level at the end of the process, while the surface temperature dropped to the air temperature (40°C).

The internal temperature distribution within the cross-section of the cylinder is shown in Fig. 11.

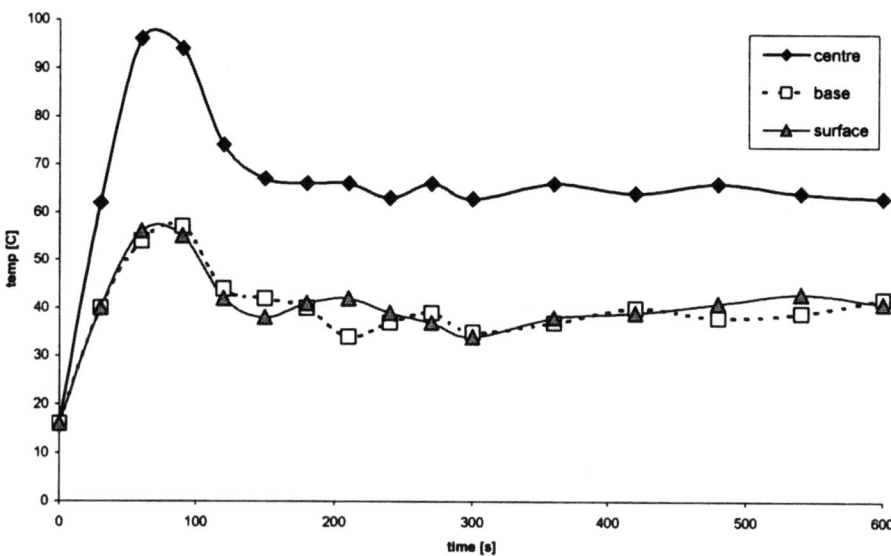

Fig. 9 Temperature distribution in the prism with triangle base

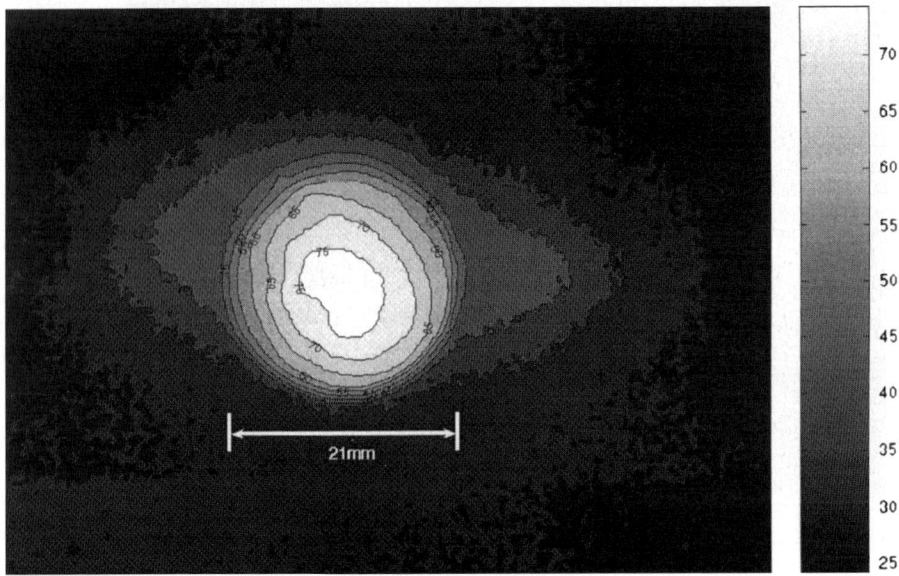

Fig. 10 IR image of the cross-section of the spherical particle (40 s of drying, temperature in Celsius degrees)

Microwave drying of the half-cylinder particle (Figs. 12 and 13) reveals smaller differences in the internal temperature distribution, but the core temperature is still higher than at the surface. At the beginning of the process the core temperature reached almost 85°C, then dropped approximately by 30°C to the level of 50°C. The

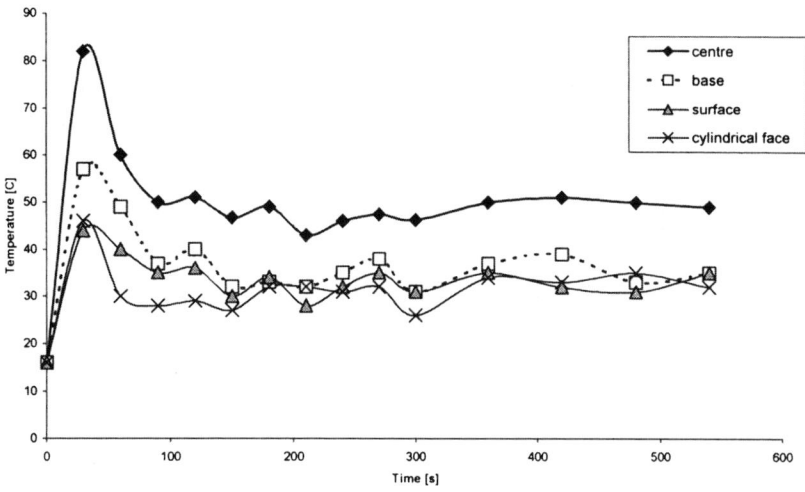

Fig. 11 IR image of the cross-section of the cylindrical particle (120 s of drying, temperature in Celsius degrees)

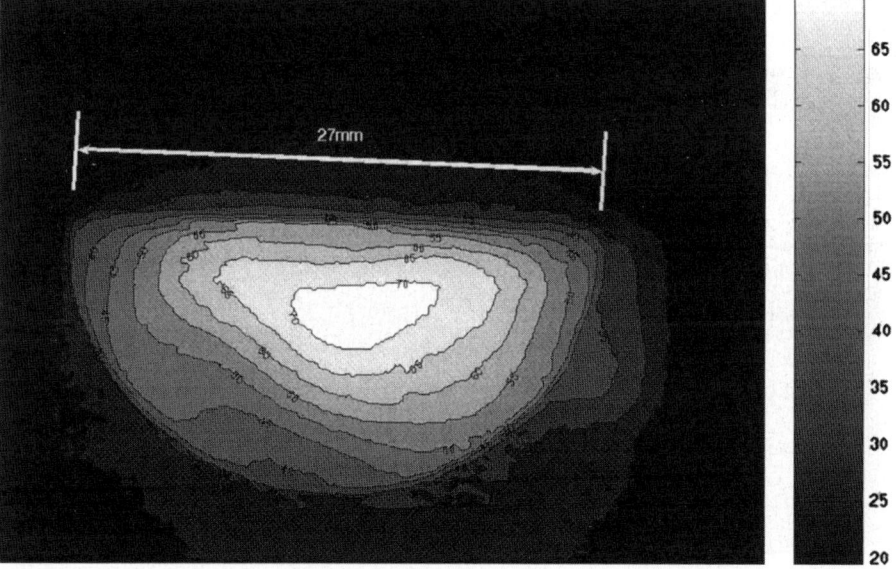

Fig. 12 Temperature distribution in half-cylindrical particle

base temperature rose faster than the other external temperatures at the beginning of the process. This difference fades after the first 100 s of drying.

The next two cases, rectangular prism (Figs. 14 and 15) and prism (Figs. 9 and 16), have different temperature distribution patterns inside. The drying rate of the rectangular prism was the highest (Fig. 6). Therefore, the drying process in this case was the highest, and the efficiency of heat generation could take place only for a very short time at the beginning of the process. The temperature reached at most 47°C. There were also differences between the core, base, and flat surface temperatures. After

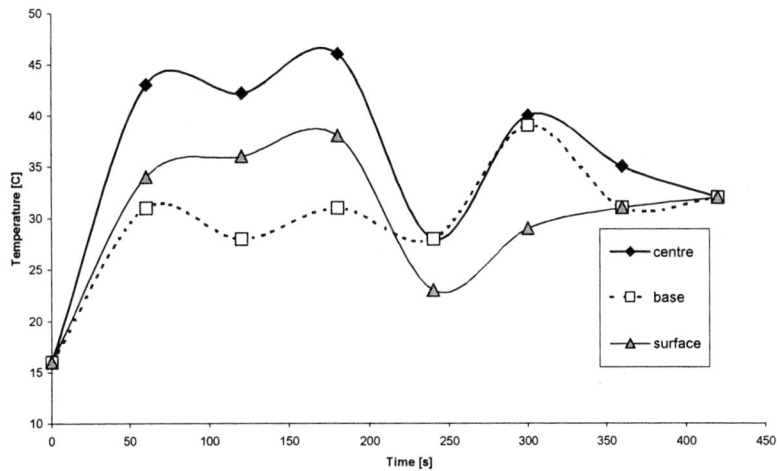

Fig. 13 IR image of the cross-section of the half-cylindrical particle (20 s of drying, temperature in Celsius degrees)

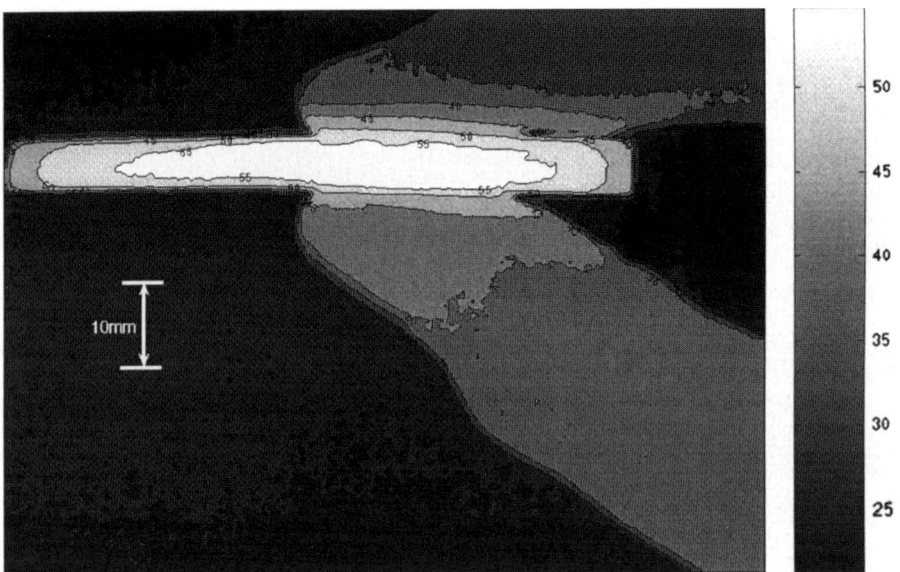

Fig. 14 Temperature distribution in rectangular prism with rectangle base

250 s of drying, the core and base temperatures dropped due to the rapid evaporation of the water from the tested material surface. Similar temperature behaviour was reported by Itaya et al. (2005).

The temperature differences inside the prism during microwave drying (Fig. 9) were the lowest in comparison to the shapes mentioned above. Even if the core temperature reached the 80°C level in the beginning of the process it would fall almost 30°C. After the first 100 s of drying, both the core and surface temperature were similar at a level a little higher than the surrounding air.

Microwave drying of various shape particles

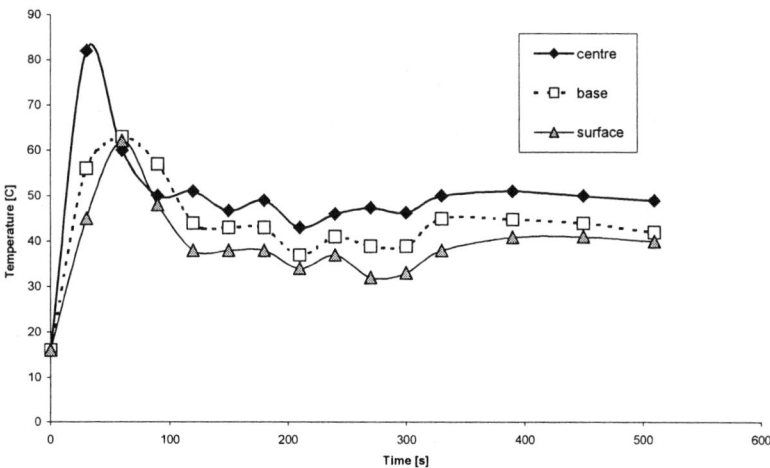

Fig. 15 IR image of the cross-section of the rectangular prism (the researcher's fingers are also visible, 180 s of drying, temperature in Celsius degrees)

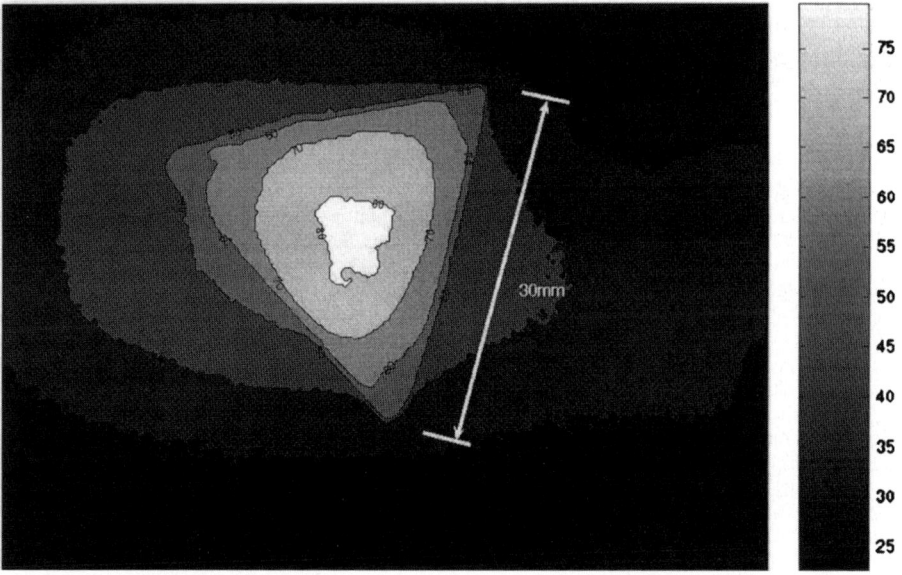

Fig. 16 IR image in the cross-section of the prism (90 s of drying, temperature in Celsius degrees)

7 Conclusions

- The maximum of the drying rate and the temperature appear at the beginning of the process, independently of the shape of the particles. There was sufficient water inside the heated particles in this period to ensure a high rate of internal heat generation.

- The particles with the largest surface area have the smallest differences between the core and the surface temperatures. Therefore, in this case the internal temperature was uniform at the end of drying.
- The level of temperature rise inside the material was directly connected with the distance between the geometrical centre and the surface (e.g. Fig. 9, where the temperature of the rectangular prism hardly reached 48°C).
- The IR images of the cross-sections of the tested particles confirmed the results obtained with the thermocouples with mean error of 2°C. The general conclusion from the IR pictures was that the most intense heat generation took place in the centre of the material independently of its shape. More detailed tests and analysis of the thermal pictures are in progress and the results will be published shortly.
- The irregularity of the temperature distribution within the material in the IR pictures is supposed to originate from both the non-uniform local moisture content and the irregularity of the solid matrix structure.
- The experimental results presented in this work may be important in case of drying of thermally sensitive materials. The drying of the sphere seems to be the most risky (the internal temperature differences were the highest in that case) in contrast to the rectangular prism shape.

Acknowledgements This work was carried out as a part of the research project No 3 T09C05826 sponsored by the Polish State Committee for Scientific Research.

References

Araszkiewicz, M., Koziol, A., Lupinski, A., Lupinski, M.: Microwave drying of porous materials. Drying Technol. **22**(10), 2331–2341 (2004)

Araszkiewicz, M., Koziol, A., Kawala, Z.: Suszenie mikrofalowe nasion rzepaku w zlozu fluidalnym (Microwave drying of the rape seeds in the fluid bed). Inzynieria Chemiczna i Procesowa, **24**, 281–291(in Polish) (2003)

Datta, A.: Heat and mass transfer in the microwave processing of food. Chem. Eng. Prog. **86**(6), 47–53 (1990)

Dincov, D.D., Parrot, K.A., Pericleous, K.A.: Heat and mass transfer in two-phase porous materials under intensive microwave heating. J. Food Eng. **65**, 403–412 (2004)

Itaya, Y., Uchiyama, S., Hatano, S., Mori, S.: Drying enhancement of clay slab by microwave heating. Drying Technol. **23**(6), 1243–1255 (2005)

Koziol, A., Araszkiewicz, M., Lupinski, A., Lupinski, M.: Microwave drying of a single porous particle suspended within fluidized bed, 7th World Congress of Chemical Engineering, 10–14 July, Glasgow, full text available on the CD.

Metaxas, A.C., Meredith, R.J.: Industrial microwave heating, Peter Peregrinus Ltd., London, UK (1983)

Meredith, R.J.: Engineers' handbook of industrial microwave heating, The Institution of Electrical Engineers, London, UK, pp. 2 (1998)

Remmen, H.H.J., Ponne, C.T., Nijhuis, H.H., Bartels, P.V., Kerkhof, P.J.A.M.: Microwave heating distributions in slabs, spheres and cylinders with relation to food processing. J. Food Sci. **61**(6), 1105–1113 (1996)

Schlunder, E.U.: Microwave drying of ceramic spheres and cylinders. Trans IchemE **71**(A), 622–628 (1993)

Wang, Z.H., Chen, G.: Theoretical study of fluidized-bed drying with microwave heating. Ind. Eng. Chem. Res. **39**, 775–782 (2000)

Periodic fluctuations of flow and porosity in spouted beds

Roman G. Szafran · Andrzej Kmiec

Received: 20 October 2005 / Accepted: 26 March 2006 /
Published online: 30 August 2006
© Springer Science+Business Media B.V. 2006

Abstract The results of experimental investigations of bed flow hydrodynamics in spouted beds are compared with CFD simulations (Eulerian–Eulerian approach) for two different column geometries. The experimental results of bed porosity and fluctuation frequency of mass flow rate of grain in the fountain region are compared with the corresponding results of simulations. The simulation results confirmed the observations of Muir et al. (1990, Chem. Eng. Comm. **88**: 153–171) and Yang and Keairns (1978, AlchE Symp. Ser. No. 176 **74**: 218) that fluctuations of bed flow in DTSB are caused by particle cluster formation in the loading region at the bottom of column. The solids cross into the jet and cover the column inlet and are carried upward periodically through a draft tube. Subsequent figures obtained from simulations, which show stages of particle cluster formation at the entrance of column, exactly match visual observations. The frequency of fluctuations of grain mass flow rate predicted in simulations ($\sim$5–6 Hz) is in the range of that experimentally determined. The fluctuating inflow of solids results in slug formation and explains the vertical variations of height and porosity of the fountain.

Keywords Porosity fluctuations · Spouted beds · CFD · Computational fluid dynamics · Drying

Notation
C_D Drag coefficient
d_s Particle diameter (m)
g Gravity acceleration (m/s^2)
h_{bed} Initial height of bed (m)
$\overline{\overline{I}}$ Identity tensor

R.G. Szafran (✉) · A. Kmiec
Department of Chemical Engineering W3/Z7,
Wroclaw University of Technology,
ul. Norwida 4/6, 50-373 Wroclaw, Poland
e-mail: roman.szafran@pwr.wroc.pl

$K_{ip}, K_{is}, K_{ps}, K_{sp}$	Interphase exchange coefficients (kg/m³ s)
$\dot{m}_{ji}$	Phase exchange term (kg/m³ s)
P	Pressure (Pa)
P_s	Solids pressure (Pa)
Re	Reynolds number
t	Time (s)
$\vec{v}$	Velocity (m/s)

Greek letters

λ	Bulk viscosity (Pa s)
η	Total fluid viscosity (Pa s)
ρ	Physical density (kg/m³)
μ	Shear viscosity (Pa s)
$\bar{\bar{\tau}}$	Stress tensor (Pa)
ω	Volume fraction

Subscripts

i, j	Different phases
p	Fluid phase
ps	Between fluid and solid phase
s	Granular phase

1 Introduction

Internally Circulating Fluidized Beds (ICFB), including Spouted Beds with a Draft Tube (DTSB) and Spout-Fluid Beds (SFB), are mainly two-phase, gas–solid or liquid–solid systems. Some special applications may also have gas–liquid–solid three-phase ICFB systems. The DTSB were initially developed for the drying of wheat grains (Mathur and Epstein 1974). Nowadays they are used for drying grain agricultural products (Devahastin et al. 1998; Kudra et al. 2001), dispersed materials (Kmiec and Szafran 2000), paste-like materials on inert bodies (Kudra et al. 1989; Arsenijevic et al. 2004), slurries and solutions (Tia et al. 1995). The SB dryers have also been used for coating particles to obtain products with special properties, such as tablets (Kucharski and Kmiec 1983; Publio and Oliveira 2004). A special area of application of this type of equipment in the pharmaceutical industry is the coating operation for producing functional particles of around 30 μm, which can be used, for example, as injectable suspensions for cancer therapy (Arimoto et al. 2004). The ICFBs are widely applied as reactors in such processes as heterogeneous catalysis (Al-Mayman and Al-Zahrani 2003), gasification of biomass (Kersten et al. 2003), coal and the pyrolysis of plastics (Aguado et al. 2003). New possible areas of applications are desulfurization of flue gases and thermal utilization of wastes (Mukadi et al. 1999).

A basic disadvantage of an ICFB apparatus is related to the scale-up and development of an industrial unit. The design of process and apparatus may be facilitated by application of such tools as CAD, CAM and CAE (Szafran and Kmiec 2004). In addition, Computational Fluid Dynamics (CFD) can help to shorten product and process development cycles and optimize processes to improve energy efficiency and environmental performance (Szafran and Kmiec 2004a). The basic problem is to establish dependencies between geometric configuration and operational parameters and the

rate of a given process and its efficiency. In the abovementioned systems the rate of the process is usually related to the heat- and mass-transfer rate between phases, which depends mainly on the hydrodynamics of bed-flow and the porosity of the bed. The hydrodynamic characteristics of ICFB systems have been investigated by a number of researchers, including Olazar et al. (2001) and San Jose et al. (2005), who analyzed spouted bed hydrodynamics.

The bed-flow hydrodynamics of a DTSB is characterized by a special regime that does not occur in any other fluidized or spouted beds. One observes periodic pulsations of bed-flow at frequencies ranging from 4 to 12 Hz. This phenomenon was observed by Muir et al. (1990), who found that for their apparatus configuration the flow of particles occurred at frequencies ranging from 8 to 12 Hz (increasing with increasing gas flow) over the range of flows studied. Moreover, they observed that the solids crossed into the jet and covered the inlet. These solids were carried away by the spouting gas, thus forming a particle cluster. Yang and Keairns (1978) also observed the formation of slugs at the draft tube inlet and studied its frequency employing high-speed film. They found that the frequency was between 4 and 8 Hz. In their study, Ji et al. (1997) examined the effect of gas velocity on the solid circulation behaviour in a DTSB. Solids circulation and motion of clusters were characterized by the spectral analyzes of the time series data of pressure fluctuation and the optical signals measured by an optical fibre probe. It was found that at low gas velocity, solid particles in the entrainment region moved from the annulus into the gas jet to form clusters periodically, and were carried upward in the draft tube with a frequency of 7–10 Hz. At higher gas velocity, solids flow was observed to become homogeneous and dilute without the formation of clusters, leading to stable pneumatic conveying of solids. Ji et al. (1998) also confirmed these observations in further investigations. Among many available mathematical models of DTSB hydrodynamics, for example that of Kmiec and Ludwig (1998), it is symptomatic that none of them takes account of periodic fluctuations of bed-flow.

The aim of this work is to verify whether a CFD modelling technique and a Eulerian–Eulerian modelling approach are capable of predicting periodic fluctuations of bed-flow in DTSBs with sufficient accuracy for engineering calculations to provide a profound analysis of bed-flow fluctuations. The results of experimental investigations are compared with CFD simulations for two different geometries of column. Experimental results of bed porosity and fluctuation frequency of mass flow rate of grain in a fountain region are compared with results of simulations.

2 Mathematical model description

The Eulerian multiphase model available in FLUENT 6.1 was chosen to carry out computer simulations. It requires solving the time averaged continuity and conservation equations for each phase, which are presented in Table 1. The volume fraction of each phase is calculated from continuity equation (1) along with the condition that the volume fractions of phases sum to one. Equations 2 and 3 describe momentum balances for fluid and granular phases, respectively. Coupling is achieved through the pressure and interphase exchange terms—the last terms on the right-hand side of Eqs. 2 and 3, where K_{ip} and K_{is} are interphase exchange coefficients. Gidaspow's model (Gidaspow 1994)—Table 2, used to describe interactions between air and solids—is a combination of the Wen and Yu (1966) model and the Ergun equation. This

Table 1 Equations of continuity and conservation

Name	Expression	
Continuity equation	$\frac{\partial}{\partial t}(\rho_i \omega_i) + \nabla \cdot (\rho_i \omega_i \vec{v}_i) = 0$	(1)
Conservation of momentum for fluid phase	$\frac{\partial}{\partial t}(\omega_p \rho_p \vec{v}_p) + \nabla \cdot (\omega_p \rho_p \vec{v}_p \vec{v}_p) = -\omega_p \nabla P + \nabla \cdot \bar{\bar{\tau}}_p + \omega_p \rho_p \vec{g}$ $+ \sum_{i=1}^{n}(K_{ip}(\vec{v}_i - \vec{v}_p) + \dot{m}_{ip} \vec{v}_{ip})$	(2)
Conservation of momentum for granular phase	$\frac{\partial}{\partial t}(\omega_s \rho_s \vec{v}_s) + \nabla \cdot (\omega_s \rho_s \vec{v}_s \vec{v}_s) = -\omega_s \nabla P - \nabla P_s + \nabla \cdot \bar{\bar{\tau}}_s + \omega_s \rho_s \vec{g}$ $+ \sum_{i=1}^{n}(K_{is}(\vec{v}_i - \vec{v}_s) + \dot{m}_{is} \vec{v}_{is})$	(3)

Table 2 Equations of interphase exchange coefficient (Gidaspow 1992)

Name	Expression	
Interphase momentum exchange coefficient for $\omega_p > 0.8$	$K_{ps} = K_{sp} = \frac{3}{4} C_{D,sp} \frac{\omega_s \omega_p \rho_p |\vec{v}_s - \vec{v}_p|}{d_s} \omega_p^{-2.65}$	(4)
Drag coefficient for $\omega_p > 0.8$	$C_{D,sp} = \frac{24}{\omega_p Re_s}\left[1 + 0.15 (\omega_p Re_s)^{0.687}\right]$	(5)
Interphase exchange coefficient for $\omega_p \leq 0.8$	$K_{ps} = K_{sp} = 150 \frac{\omega_s(1-\omega_p)\mu_p}{\omega_p d_s^2} + 1.75 \frac{\rho_p \omega_s |\vec{v}_s - \vec{v}_p|}{d_s}$	(6)
Relative Reynolds number	$Re_s = \frac{\rho_p |\vec{v}_s - \vec{v}_p| d_s}{\mu_p}$	(7)

model is appropriate for dilute and dense granular flow regimes and was investigated for fluidized beds. P_s is a solids pressure, which represents the force resulting from particle–particle interactions. For granular flows in the compressible regime (i.e., where the solids volume fraction is less than its maximum allowed value), a solids pressure was calculated independently and used for the pressure gradient term in the granular-phase momentum equation (3).

Eulerian models require additional closure laws to describe the rheology of particle phase and turbulence equations, which describe effects of turbulent fluctuations of velocities on scalar quantities. To describe the flow behaviour of a fluid–solid mixture the multi-fluid Gidaspow model (Gidaspow et al. 1992) and the k-ε turbulence model (Launder and Spalding 1972) were applied.

3 Computational fluid Dynamics simulations

Simulations were conducted for two different geometries: our own apparatus and apparatus of Ji et al. (1998). The main construction parameters of both apparatuses are collected in Table 3.

Table 3 Main parameters of apparatus used in simulations

Parameter Apparatus	Value DTSB
Our construction of apparatus	
Total height of column (m)	0.85
Height of cylindrical part of column (m)	0.7
Column diameter (m)	0.17
Cone angle (degree)	50
Column diameter at the inlet (m)	0.03
Column diameter at the outlet (m)	0.1
Height of the draft tube (m)	0.116
Diameter of the draft tube (m)	0.029
Distance between the inlet of column and the inlet of draft tube (m)	0.068
Apparatus of Ji et al. (1998)	
Total height of column (m)	1.825
Height of cylindrical part of column (m)	1.525
Column diameter (m)	0.205
Cone angle (degree)	60
Column diameter at the inlet (m)	0.031
Column diameter at the outlet (m)	0.1
Height of the draft tube (m)	0.8
Diameter of the draft tube (m)	0.05
Distance between the inlet of column and the inlet of draft tube (m)	0.075

The control-volume-based code FLUENT 6.1 was chosen to carry out computer simulations. Because of the system's symmetry, to reduce computational times and system resources required, the 2D axisymmetric segregated solver was chosen and simulations of only one half-domain were carried out. The computational meshes composed of triangular cells in the core and quadrilateral cells in the near wall region are shown in Fig. 1 and their parameters are collected in Table 4. The mesh density was chosen to fulfil conditions for near-wall function and to minimize the solution's dependence on mesh density. As Fig. 1 shows, the density of meshes increases near the inlet, in the annulus region and in a draft tube to allow proper prediction of two-phase flow — particle cluster formation.

Pressure-velocity coupling was achieved using the SIMPLE algorithm. The second-order upwind discretization scheme of momentum, volume fraction of phases, energy, user-defined scalars, turbulence kinetic energy, turbulence dissipation rate and the first-order implicit time discretization were chosen. The typical values of under-relaxation factors are 0.2–0.4. As criteria of convergence, the value of scaled residuals and volume integrals were monitored. The solution was considered converged when the scaled residuals were less than 1×10^{-3} for all variables except continuity and energy, the criterion of convergence of which was set to 1×10^{-4} and 1×10^{-6}, respectively. The maximum number of iterations per time step was set to 100. The time step in unsteady simulations varied, depending on solution convergence, between 3×10^{-4} and 1×10^{-4} s.

The following boundary conditions were applied in every simulation:

(1) Continuous phases are treated as ideal gas.
(2) The solid phase is incompressible.
(3) The flow is incompressible.

Fig. 1 Computational meshes: (**a**) our own apparatus, (**b**) Ji et al. (1998)

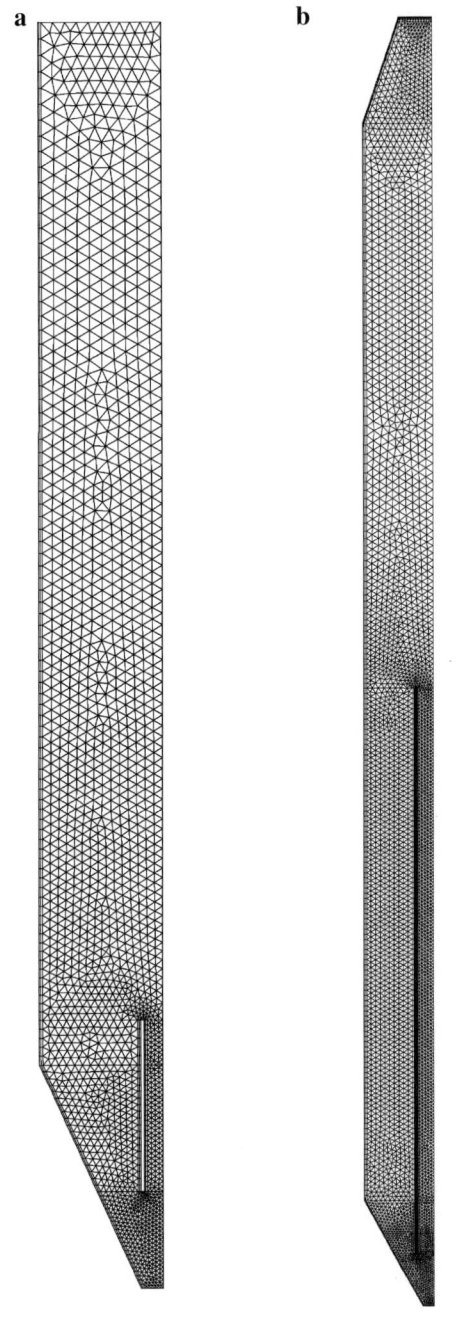

Table 4 Parameters of computational meshes

Parameter	Value
Our construction of apparatus—Fig. 1a	
Number of cells	3985
Number of nodes	2357
Minimal cell area (m^2)	5.2×10^{-7}
Maximal cell area (m^2)	5.3×10^{-5}
Mean cell area (m^2)	2.5×10^{-5}
Mean squish index	6.2×10^{-3}
Mean equiangle skew	4.9×10^{-2}
y^+	0–28[a]
Apparatus of Ji et al. (1998)—Fig. 1b	
Number of cells	8823
Number of nodes	5669
Minimal cell area (m^2)	4.4×10^{-7}
Maximal cell area (m^2)	8.6×10^{-5}
Mean cell area (m^2)	2.2×10^{-6}
Mean squish index	7.1×10^{-3}
Mean equiangle skew	6.6×10^{-2}
y^+	0–65[a]

[a] Rough value

(4) The walls are adiabatic and non-slip wall conditions were used for all phases.
(5) The inlet condition was considered as velocity inlet boundary condition with uniform velocity profile.
(6) The pressure boundary condition was considered at the outlet.
(7) The axis symmetry boundary condition was applied along the axis of symmetry.

4 Experimental setup

We investigated the hydrodynamics of bed-flow for three kinds of grain, the properties of which are collected in Table 5. A diagram of the experimental setup is shown in Fig. 1. The dryer column 1 was 0.7 m high, the diameter of the cylindrical part was 0.17 m, inlet diameter was 0.03 m, and cone angle was 50°. Inside the column there was a fixed draft tube 2, 0.027 m I.D. and 0.116 m high at a distance of 0.068 from the bottom of the column. At the top and bottom of the column there were wire nets, which prevented grain escaping from the dryer. Gas velocity was measured by means of orifice 7. Moisture content in the air was measured by means of psychrometer 3. Valve 5 was closed during experiments (Fig. 2).

Detailed descriptions of the apparatus that was used by Ji et al. (1998) were presented in their publication. Basic parameters of the apparatus are collected in Table 3.

5 Results and discussion

Experimental investigations demonstrate that bed-flow in DTSP apparatus is characterized by a periodic character. In the fountain region, periodic fluctuations of bed porosity in each experiment were observed. In addition, for fine particles, instabilities

Table 5 Properties of grain

Parameter	Value
Our construction of apparatus	
Name of material	Microsphere
Density of grain (kg/m^3)	630
Packing density (kg/m^3)	400
Particle diameter (m)	2.2×10^{-4}
Geldart's group	A
Name of material	Rapeseed
Density of grain (kg/m^3)	1078
Packing density (kg/m^3)	678
Particle diameter (m)	2×10^{-3}
Geldart's group	D
Name of material	Polyethylene
Density of grain (kg/m^3)	871
Packing density (kg/m^3)	580
Particle diameter (m)	3.7×10^{-3}
Geldart's group	D
Apparatus of Ji et al. (1998)	
Name of material	Glass beads
Density of grain, kg/m^3	2500
Packing density, kg/m^3	1500
Particle diameter, m	1×10^{-3}
Geldart's group	D

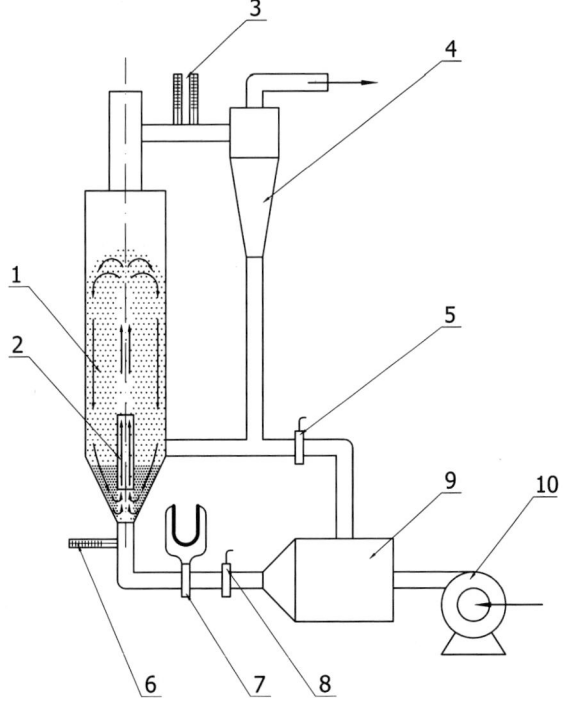

Fig. 2 Schematic diagram of experimental setup: 1 dryer, 2 draft tube, 3 dry- and wet-bulb thermometer, 4 cyclone, 5, 8 valves, 6 thermometer, 7 orifice, 9 heater, 10 fan

Fig. 3 Experimental distributions of grain (rapeseed) concentration in a fountain region at different velocities; initial height of bed 0.15 m: (**a**) 8 m/s, (**b**) 10 m/s, (**c**) 11 m/s, (**d**) 14 m/s

of bed-flow and height of fountain were noticeable. A detailed description of flow hydrodynamics of microsphere grains in a DTSB has been presented earlier by Szafran et al. (2005). In general, spouting becomes more stable for larger particle diameter and higher inlet air velocity. Figure 3 shows example distributions of grain concentration in the fountain region obtained for rapeseed at different air inlet velocities. As the inlet air velocity increases, the porosity of the fountain increases and the distribution of grain concentration becomes more homogenous. As one can see in Fig. 3a, the upper limit of the fountain is sharp for low air inlet velocities and the grain concentration reaches its maximum value at the top of the fountain. On the other hand, as one can

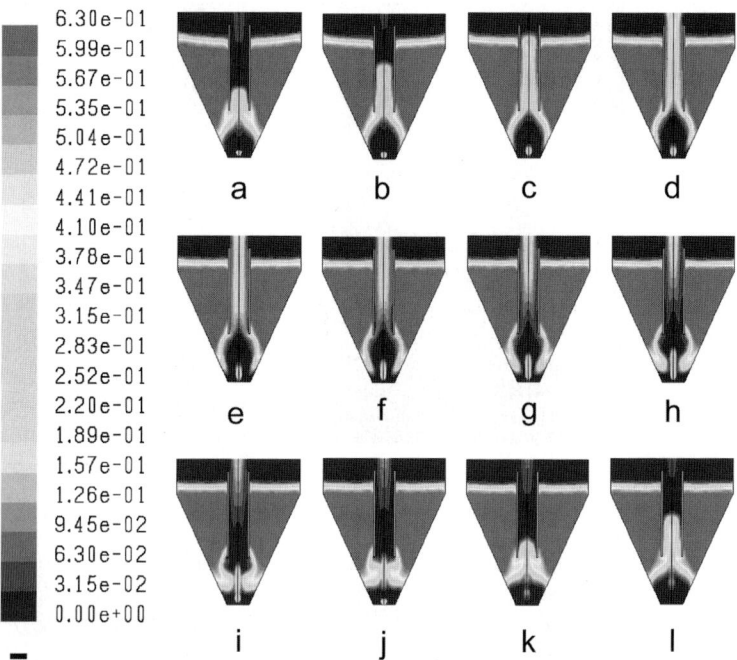

Fig. 4 The CFD simulations of bed-flow at the bottom of column for rapeseed and inlet air velocity 14 m/s. Cluster of grain formation. Time step 0.015 s. Volume fraction of *solid*

see in Fig. 3b–d, for high inlet air velocity, the upper limit of the fountain becomes fuzzier. The same relationships were observed for all kinds of grain.

The CFD simulations were performed for the same hydrodynamic conditions. Numerical simulations predict transient behaviour of flow in the system investigated. Particles in the entrainment region move from the annulus region into the gas jet to form clusters and they are carried upward periodically. This agrees with the visual observations made during our experimental investigations and has been reported by various researchers, as mentioned above. Figures 4a–l present the results of simulations of bed-flow for the bottom part of the column (annulus region). The initial conditions were the same as in the case presented in Fig. 3d. Subsequent figures present stages of cluster formation from grain at the entrance of column and its transport through the draft tube. This behaviour explains the observed periodic fluctuations of bed-flow and fountain porosity. The solids mass flow rate at the end of the draft tube changes periodically with a frequency of 6 Hz. Figure 5 presents the dependence of mass flow rate of grain at the end of the draft tube on time. Similar fluctuations, of frequency 5 Hz, were also obtained for microsphere and polyethylene grains (see Figs. 6 and 7), respectively. For polyethylene grains, the amplitude of fluctuations was nearly constant in time at about 5 kg/s. This kind of grain exhibited stable spouting in experiments. For rapeseed and microspheres, the amplitude of fluctuation changed over time, which caused instabilities in spouting and during experiments was observed as a fluctuation of fountain height and porosity.

The results of CFD simulations of spouting for the apparatus of Ji et al. (1998) indicate, as in the cases discussed above, a periodic character of grain loading at the

Fig. 5 The CFD simulations. Dependence of mass flow rate of grain at the end of draft tube on time; inlet air velocity 14 m/s; ; initial height of bed 0.155 m; rapeseed

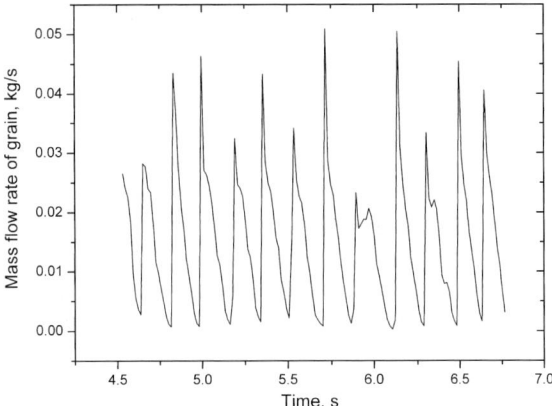

Fig. 6 The CFD simulations. Dependence of mass flow rate of grain at the end of draft tube on time; inlet air velocity 5.5 m/s; initial height of bed 0.12 m; microsphere

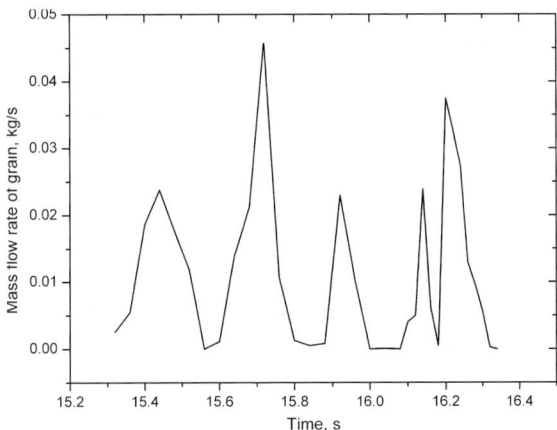

bottom of the column in the loading region and its periodic flow through the draft tube. The authors observed a similar bed-flow behaviour during their experimental investigations. Figure 8 shows the distributions of grain concentration in the column of apparatus for different inlet air velocities. In cases 8a and b, a stable fountain was obtained in which porosity reached a grain packing density at the top of the fountain. In case 8c, grain was blown away from the column, so the fountain was not obtained. The authors did not confirm this event in experiments. In Figure 9, the results of simulations of fluctuation frequency of grain mass flow rate are compared with experimental data. As one can see, simulations predict too low values of fluctuation frequency for low inlet air velocities, but the predicted tendency is right. The frequency increases as inlet air velocity increases.

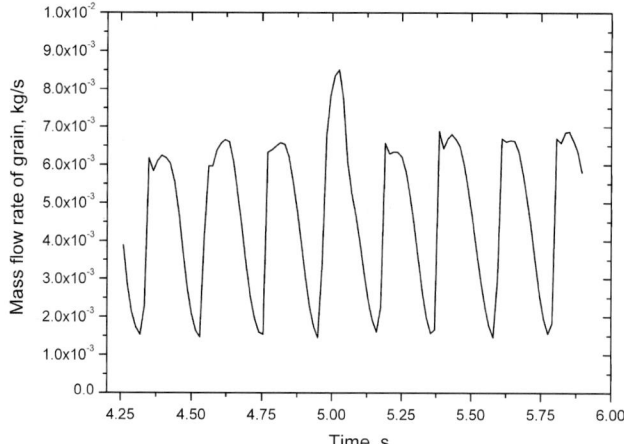

Fig. 7 The CFD simulations. Dependence of mass flow rate of grain at the end of draft tube on time; inlet air velocity 14 m/s; initial height of bed 0.13 m; polyethylene

6 Conclusions

The results of simulations confirmed the observations of Muir et al. (1990) and Yang and Keairns (1978) that fluctuations of bed flow in DTSB are caused by particle cluster formation in the loading region at the bottom of column. The solids cross into the jet and cover the inlet of the column and are carried upward periodically through a draft tube. Subsequent figures obtained from simulations, which show stages of particle cluster formation at the entrance of column, exactly match visual observations reported by Muir et al. (1990). The frequency of fluctuations of grain mass flow rate predicted in simulations ($\sim$ 5–6 Hz) are in the range of that determined experimentally by Yang and Keairns (1978) and close to the range reported by Ji et al. (1998) of 7–10 Hz and by Muir et al. (1990) of 8–12 Hz. The fluctuating inflow of solids results in the slug formation and explains the vertical variations of height and the porosity of the fountain.

References

Aguado, R., Olazar, M., Gaisan, B., Prieto, R., Bilbao, J.: Polystyrene pyrolysis in a conical spouted bed reactor. Chem. Eng. J. **92**, 91–99 (2003)

Al-Mayman, S.I., Al-Zahrani, S.M.: Catalytic cracking of gas oils in electromagnetic fields: Reactor design and performance. Fuel Process. Technol. **80**, 169–182 (2003)

Arimoto, M., Ichikawa, H., Fukumori, Y.: Microencapsulation of water-soluble macromolecules with acrylic terpolymers by the Wurster coating process for colon-specific drug delivery. Powder Technol. **141**, 177–186 (2004)

Arsenijevic, Z.Lj., Grbavcic, Z.B., Garic-Grulovic, R.V.: Drying of suspensions in the draft tube spouted bed. Can. J. Chem. Eng. **82**(3), 450–464 (2004)

Devahastin, S., Mujumdar, A.S., Raghavan, G.S.V.: Diffusion-controlled batch drying of particles in a novel rotating jet annular spouted bed. Dry. Technol. **16**(3–5), 525–543 (1998)

Gidaspow, D., Bezburuah, R., Ding, J.: Hydrodynamics of circulating fluidized beds, Kinetic theory approach. In: Fluidization VII, Proceedings of the 7th Engineering Foundation Conference on Fluidization, pp. 75–82.

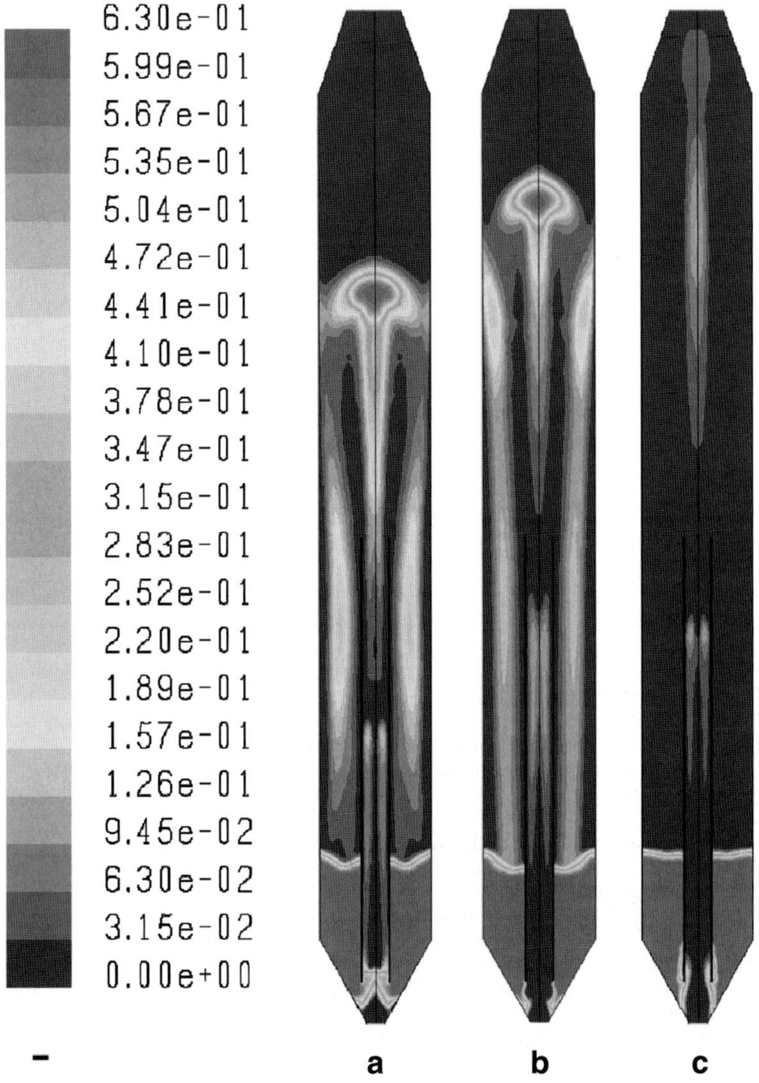

Fig. 8 The CFD simulations. Distribution of grain concentration in the apparatus for different inlet air velocities: (**a**) 35 m/s, (**b**) 50 m/s, (**c**) 70 m/s. Volume fraction of *solid*

Gidaspow, D.: Multiphase Flow and Fluidization. Academic Press Inc., San Diego, CA (1994)
Ji, H., Tsutsumi, A., Yoshida, K.: Characteristics of particle circulation in a spouted bed with a draft tube. AIChE Symp. Ser. No. 317, **93**, 131–135 (1997)
Ji, H., Tsutsumi, A., Yoshida, K.: Solid circulation in a spouted bed with a draft tube. J. Chem. Eng. Jpn, **5**(31), 842–845 (1998)
Kersten, S.R.A., Prins, W., van der Drift, B., van Swaaij, W.P.M.: Principles of a novel multistage circulating fluidized bed reactor for biomass gasification. Chem. Eng. Sci. **58**, 725–731 (2003)
Kmiec A., Szafran, R.G.: Kinetics of drying of microspherical particles in a spouted bed dryer with a draft tube. In: Proceedings of the 12th International Drying Symposium (IDS 2000); Elsevier Science B.V., Amsterdam (2000)

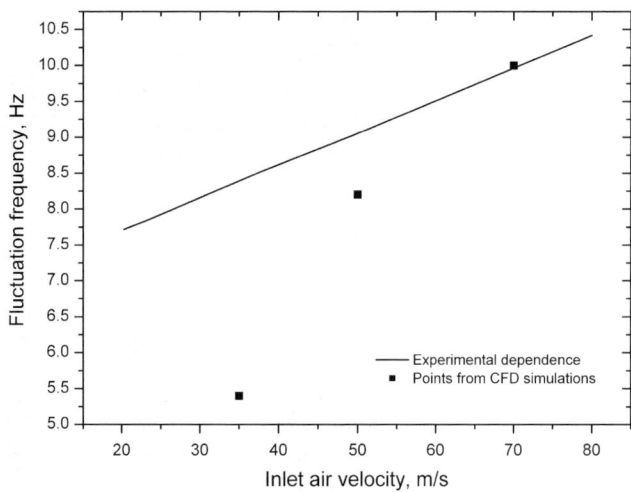

Fig. 9 Comparison of results of simulations with experimental data. Dependence of fluctuation frequency of grain mass flow rate on inlet air velocity

Kmiec, A., Ludwig, W.: Model przeplywu dwufazowego gaz—cialo stale w aparacie fontannowym z rura wznoszaca. I Wyprowadzenie modelu. Inz Chem. Procesowa **19**, 575–589 (1998)

Kucharski, J., Kmiec, A.: Hydrodynamics, heat and mass transfer during coating of tablets in a spouted bed. Can. J. Chem. Eng. **61**, 435–439 (1983)

Kudra, T., Pallai, Z., Bartczak, Z., Peter, M.: Drying of paste-like materials in screw-type spouted-bed and spin-flash dryers. Drying Technol. **7**, 583–597 (1989)

Kundu, K., Datta, A., Chatterjee, P.: Drying of oilseeds. Dry. Technol. **19**, 343–358 (2001)

Launder, B.E., Spalding, D.B.: Lectures in Mathematical Models of Turbulence. Academic Press, London (1972)

Mathur, K.B., Epstein, N.: Spouted Beds. Academic Press Inc., New York (1974)

Muir, J.R., Berruti, F., Behie, L.A.: Solids circulation in spouted and spout-fluid beds with draft-tubes. Chem. Eng. Comm. **88**, 153–171 (1990)

Mukadi, L., Guy, Ch., Legros, R.: Modeling of an internally circulating fluidized bed reactor for thermal treatment of industrial solid wastes. Can. J. Chem. Eng. **77**, 420–431 (1999)

Olazar, M., san Jose, M.J., Izquierdo, M.A., Ortiz de Salazar, A., Bilbao, J.: Effect of operating conditions on solid velocity in the spout, annulus and fountain of spouted beds. Chem. Eng. Sci. **56**, 3585–3594 (2001)

Publio, M.C.P., Oliveira, W.P.: Effect of the equipment configuration and operating conditions on process performance and on physical characteristics of the product during coating in spouted bed. Can. J. Chem. Eng. **82**(1), 122–133 (2004)

San Jose, M.J., Olazar, M., Alvares, S., Morales, A., Bilbao, J.: Local porosity in conical spouted beds consisting of solids of varying density. Chem. Eng. Sci. **60**, 2017–2025 (2005)

Szafran, R.G., Kmiec, A.: Using of computer techniques in design process of disperse gas–solid systems. Inz. Chem. Procesowa **25**, 1645–1650 (in Polish) (2004)

Szafran, R.G., Kmiec A.: CFD modeling of heat and mass transfer in a spouted bed dryer. Ind. Eng. Chem. Res. **43**, 1113–1124 (2004a)

Szafran, R.G., Kmiec, A., Ludwig, W.: CFD modeling of a spouted-bed dryer hydrodynamics. Dry. Technol. **23**(8), 1723–1736 (2005)

Tia, S., Tangsatikulchai, C., Dumronglaohapun, P.: Continuous drying of slurry in a jet spouted bed. Dry. Technol. **13**, 1825–1840 (1995)

Wen, C.Y., Yu, Y. H.: Mechanics of fluidization. Chem. Eng. Prog. Symp. Ser. **62**, 100–111 (1966)

Yang, W.C., Keairns, D.L.: Design of recirculating fluidized beds for commercial applications. AIChE Symp. Ser. No. 176. **74**, 218 (1978)

Modeling of vacuum desorption of multicomponent moisture in freeze drying

J. F. Nastaj · B. Ambrożek

Received: 22 October 2005 / Accepted: 26 March 2006 /
Published online: 30 August 2006
© Springer Science+Business Media B.V. 2006

Abstract A mathematical model of multicomponent vacuum desorption, which occurs in the vacuum freeze drying process has been developed. Drying with conductive heating and constant contact surface temperature was considered. Pressure drop in the layer of the material to be dried was taken into account in the model formulation and process simulation. Equilibrium moisture content for pure water, toluene, and m-xylene and their two- and three-component mixtures on zeolite DAY 20F were described by means of the multitemperature extended Langmuir isotherm equation. Model equations were solved by the numerical method of lines. Moisture content and temperature distributions within the drying material were predicted from the model as a function of drying time.

Keywords Vacuum desorption · Freeze drying · Multicomponent moisture · Zeolite DAY F20

Nomenclature

a_e Effective thermal diffusivity, m²/s.
a_i Constant of multitemperature Langmuir isotherm of component i, K.
b_i Constant of multitemperature Langmuir isotherm of component i, Pa^{-1}.
c_{a_i} Specific heat of component i in adsorbed phase, J/(kg K).
c_{pg_i} Specific heat of component i in gas (vapor) phase, J/(mol K).
c_i Constant of multitemperature Langmuir isotherm of component i, K.
c_s Specific heat of adsorbent, J/(kg K).
C Molar gas density, mol/m³.
C_{i0} Initial molar concentration of component i in the gaseous phase, mol/m³.
C_i^* Dimensionless molar concentration of component i in the gas phase.

J. F. Nastaj (✉) · B. Ambrożek
Department of Chemical Engineering and Environmental Protection Processes, Szczecin University of Technology, Al. Piastów 42, 71-065 Szczecin, Poland
e-mail: jonas@ps.pl

B. Ambrożek
e-mail: ambog@ps.pl

d_p Particle diameter, m.
D_{eff_i} Effective diffusion coefficient of component i, m²/s.
D_{K_i} Knudsen diffusion coefficient of component i, m²/s.
D_{S_i} Surface diffusion coefficient of component i, m²/s.
ΔH_i Heat of adsorption of component i, J/mol.
J_z Mass flux density, mol/(m² s).
k_e Effective thermal conductivity of adsorbent bed, W/(mK).
k_D Permeability of porous media (m²).
k_E Parameter describing the inertial effect in Ergun's equation, m.
K_i Kinetic coefficient of component i, 1/s.
L Dried layer thickness, m.
M_i Molecular weight of component i, kg/mol.
N Number of sections of drying material layer.
p_i Partial pressure of component i, Pa.
ΔP Reference pressure variation for normalization, Pa.
P Total pressure, Pa.
X_{avg_i} Average moisture content of component i (dry basis), kg i/kg.
X_i Moisture content of component i (dry basis), kg i/kg.
X_i^* Equilibrium moisture content of component i (dry basis), kg i/kg.
y_i Mole fraction of component i in the gas phase, mol i/mol.
t Time, s.
ΔT Reference temperature variation for normalization, K.
T Temperature, K.
T_i Initial temperature of drying layer, K.
T_0 Contact surface temperature, K.
y_i Mole fraction of component i, mol i/mol.
z Axial coordinate, m.
Z Dimensionless axial coordinate.

Greek symbols
ε Bed void fraction.
ε_p Particle porosity.
μ Viscosity of gas mixture, Pa s.
ω Normalized pressure.
θ Normalized temperature.
Ω_i Dimensionless concentration of component i in solid phase.
ρ_p Particle density of adsorbent, kg/m³.
ρ_b Bulk density of adsorbent, kg/m³.
τ Dimensionless time.

Acronyms
VC Vacuum chamber.
LDF Linear driving force.

1 Introduction

In our earlier work (Nastaj and Ambrożek 2005), we derived a simplified mathematical model of vacuum desorption in the freeze-drying process for the system isopropanol

Fig. 1 Physical model of vacuum desorption in the secondary freeze-drying and division of dried material layer into N sections

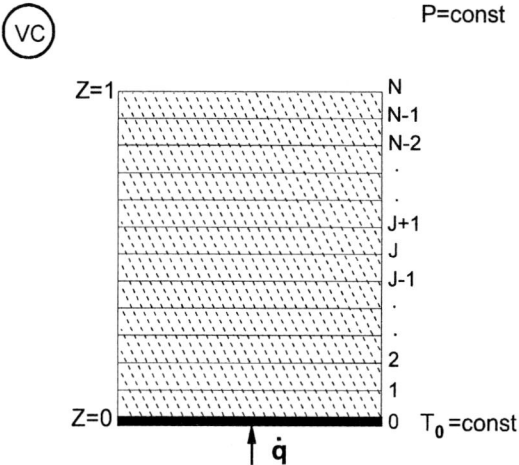

(1)—water (2)—zeolite DAY-55. The solution of that numerical model was based on two assumptions: first, that the pressure drop in the material is neglected due to the thin drying layers in vacuum freeze drying and the low vapor mass flux densities that occur. Second, that the vacuum pump immediately removes vapors coming from the material, so the vapor concentrations of the components are practically equal to zero. This paper develops the full model of a multi-component vacuum desorption, taking into account both pressure drop in the drying layer and the existence of the component vapor concentrations inside the material. This should confirm or reject the correctness of the assumptions made in our preliminary investigations.

The freeze-drying process is conventionally divided into three stages: pre-freezing, sublimation drying under vacuum (known as primary drying), followed by desorption (known as secondary drying). After primary drying, residual moisture content may be as high as 7% (Rowe and Snowman 1978). Secondary drying is intended to reduce this to an optimum value for material stability, usually with moisture content between 0.5 and 2.0%. Moisture desorption follows the primary drying in a vacuum chamber, usually between room temperature (20°C) and 40°C (Goldblith et al. 1975).

Secondary drying is the period of the removal of moisture contained in micropores of the material being dried. Removal of multicomponent moisture is a common problem in pharmaceutical, biotechnology and related industries (Laurent et al. 1999; Wolf and Schlünder 1999).

A schematic description of the secondary freeze-drying process during contact heating is presented in Fig. 1, which additionally presents the division of the material layer into N sections with $N + 1$ node points according to the numerical method of lines, which will be used later. Many papers concerning mathematical modeling of vacuum freeze drying have already been published.

Studies dealing with mathematical modeling of the primary and secondary stages of bulk solution freeze drying in trays (Sadikoglu and Liapis 1997) and pharmaceutical crystalline and amorphous solutes in vials (Liapis and Bruttini 1995) have been carried out.

Experimental and theoretical research and development studies in important areas of the freeze-drying process were presented in Liapis et al. (1996). However, most of these studies are limited to desorption of single component moisture, in the secondary

Table 1 Characteristics of Zeolite DAY 20F adsorbent

Adsorbent	Zeolite DAY 20F
Surface area (BET) (m^2/g)	704
Equivalent particle diameter (m)	$5 \cdot 10^{-4}$
Bulk density (kg/m^3)	500
Particle porosity (−)	0.415
Bed porosity (−)	0.677
Mean pore radius (nm)	2.17
Specific heat (J/kg K)	900

Table 2 Basic properties of adsorbates

Property	Water (1)	Toluene (2)	m-Xylene (3)
Formula	H_2O	C_7H_8	C_8H_{10}
Molar mass (kg/kmol)	18.015	92.141	106.167
Boiling point (K)	373.15	383.79	412.34
Dipole moment (Debye)	1.8	0.4	0.3

stage of freeze drying, usually water. There is a gap in the mathematical model of vacuum desorption of multicomponent moisture in freeze drying.

Zeolite DAY 20F was used as adsorbent and model material to be dried. Table 1 shows the properties of zeolite DAY 20F (Kim et al. 2005).

The basic properties of adsorbates considered in the modeling of the vacuum desorption process, which formed both single-component and multicomponent moisture are presented in Table 2 (Poling et al. 2000).

The aim of this study is to analyze vacuum desorption of one-, two-, and three-component moisture for the system water (1)—toluene (2)—m-xylene (3) adsorbed on zeolite DAY 20F, and to determine drying times for secondary vacuum freeze drying.

2 Mathematical model of fixed bed vacuum desorption

The model equations for multicomponent vacuum desorption in a fixed bed consist of the material balances of the gas and solid phase and a quasi-homogeneous energy balance. Axial dispersion is neglected but heat conduction is taken into account because of contact heating of the sample in the process of vacuum desorption.

The material balances for the all components in the gas phase can be written as:

$$\frac{\partial(y_i \cdot C)}{\partial t} + \frac{\partial(y_i \cdot J_z)}{\partial z} + \frac{\rho_b}{\varepsilon M_i} \frac{\partial X_i}{\partial t} = 0 \qquad (1)$$

with $\sum_{i=1}^{s} y_i = 1$; s is the number of all components in the gas phase including air, which is not an adsorbed component, and $C = P/RT$, which is the molar gas density.

For the adsorbed phase, a driving force approach was used. Thus:

$$\frac{\partial X_i}{\partial t} = K_i(X_i^* - X_i) \qquad (2)$$

The kinetic coefficient of component i, K_i, was calculated according to the linear driving force (LDF) approach (Glueckauf 1955):

$$K_i = \frac{60 D_{eff_i}}{d_p^2} \quad (3)$$

or the quadratic driving force (QDF) approach of Vermeulen (Hall et al. 1966):

$$K_i = \frac{4\pi^2 D_{eff_i}}{d_p^2} \frac{X_i^* + X_i}{2X_i} \quad (4)$$

In the case of vacuum desorption, the pore diffusion is the rate-controlling step (Boger et al. 1997). The effective diffusion coefficient, D_{eff_i}, was estimated from the following equation:

$$D_{eff_i} = D_{S_i} + D_{K_i} \frac{\varepsilon_p M_i}{\rho_p} \frac{\partial y_i^*}{\partial X_i} \quad (5)$$

which considers Knudsen and surface diffusion.

The quasihomogeneous heat balance relation can be written as:

$$-k_e \frac{\partial^2 T}{\partial z^2} + J_z \sum_{i=1}^{s} y_i c_{pg_i} \frac{\partial T}{\partial z} + c_\Sigma \frac{\partial T}{\partial t} + \rho_b \left(\sum_{i=1}^{s} \frac{\Delta H_i}{M_i} \right) \frac{\partial X_i}{\partial t} = 0 \quad (6)$$

where c_Σ is the overall volumetric heat capacity:

$$c_\Sigma = \rho_b \left[c_s + \sum_{i=1}^{s} (X_i c_{a_i}) \right] + \varepsilon \cdot C \sum_{i=1}^{s} y_i c_{pg_i} \quad (7)$$

Equation (6) includes the heat generated by adsorption of adsorbates. The isosteric heat of adsorption is estimated using an equation of the Clausius–Clapeyron type (Hill 1949; Do 1998) for both single- and mixed-gas adsorption:

$$\Delta H = -RT^2 \left(\frac{\partial \ln P}{\partial T} \right)_X \quad (8a)$$

$$\Delta H_i = -RT^2 \left(\frac{\partial \ln p_i}{\partial T} \right)_{X_i} \quad (8b)$$

We derived the analytical form of the isosteric heat of adsorption equation from the Clausius–Clapeyron equation and the extended Langmuir isotherms.

$$\Delta H_i = -\frac{R X_{io} \exp(a_i/T)}{X_{io} \exp(a_i/T) - X_i} \left[a_i + c_i - \frac{X_i c_i}{X_{io} \exp(a_i/T)} \right] \quad (9)$$

The axial pressure drop is taken into account and is given by Ergun's equation (Ergun 1952)

$$\frac{\partial P}{\partial z} = -\frac{\mu}{C \cdot k_D} J_z - \sum_{i=1}^{s} y_i M_i \frac{k_E}{C k_D} J_z^2 \quad (10)$$

where k_D is the permeability of the adsorbent bed and k_E is a parameter describing the inertial effect. These two parameters depend only on medium properties and, in the case of granular porous adsorbent (zeolite DAY 20F), are defined by:

$$k_D = \frac{d_p^2 \varepsilon^3}{150(1-\varepsilon)^2}; \quad k_E = \frac{1.75 d_p}{150(1-\varepsilon)} \tag{11}$$

where d_p and ε are the adsorbent particle diameter and porosity of the adsorbent bed, respectively. It should be noted that the Ergun equation can reduce to the Darcy law when k_E tends to be zero.

For computational convenience, the following explicit relation for the molar vapor flux density is derived (Sun et al. 1995):

$$J_z = -\frac{2 C k_D \mu^{-1} \partial P/\partial z}{1 + \sqrt{1 + 4C \cdot (\sum_{i-1}^{s} y_i M_i) k_D k_E \mu^{-2} |\partial P/\partial z|}} \tag{12}$$

Adsorption equilibria provide the key information for the design of practical separation processes based on adsorption mechanisms. Moreover, they are used to simulate the adsorption/desorption processes. The multicomponent adsorption equilibrium is usually predicted using single-component isotherm information (Yao 2000).

The adsorption equilibrium was described by the multitemperature extended Langmuir equation for a multicomponent mixture (Chahbani and Tondeur 2001).

$$X_i^* = X_{i0} \exp\left(\frac{a_i}{T}\right) \left[\frac{b_i \exp(c_i/T) P \cdot y_i}{1 + \sum_i b_i \exp(c_i/T) P \cdot y_i}\right] \tag{13}$$

The following temperature-dependent extended Langmuir equation explains the equilibrium for each mixture component:

$$X_1^* = X_{10} \exp\left(\frac{a_1}{T}\right) \left[\frac{b_1 \exp(c_1/T) P \cdot y_1}{1 + \sum_i b_1 \exp(c_1/T) P \cdot y_1}\right] \tag{14a}$$

$$X_2^* = X_{20} \exp\left(\frac{a_2}{T}\right) \left[\frac{b_2 \exp(c_2/T) P \cdot y_2}{1 + \sum_i b_2 \exp(c_2/T) P \cdot y_2}\right] \tag{14b}$$

$$X_3^* = X_{30} \exp\left(\frac{a_3}{T}\right) \left[\frac{b_3 \exp(c_3/T) P \cdot y_3}{1 + \sum_i b_3 \exp(c_3/T) P \cdot y_3}\right] \tag{14c}$$

The results of nonlinear estimation of the above equation coefficients for water (1)—toluene (2)—m-xylene (3) onto zeolite DAY 20F are presented in Table 3. The estimation was done on the basis of the literature data published by Ryu et al. (2002), Brihi et al. (2002) and Kim et al. (2005).

The residual value analysis shows that Langmuir equation gives an approximation with standard deviation 9.8% for water, 25.9% for toluene and 1.9 % for m-xylene.

Table 3 Isotherm constants for a temperature-dependent Langmuir equation. System: Water (1)—Toluene (2)—m-Xylene (3) onto Zeolite DAY 20F

Component i		Langmuir isotherm constants			
		X_{i0} (kg i/kg)	a_i (K)	b_i (Pa^{-1})	c_i (K)
1	Water	0.370300	1387.82	0.170·10^{-6}	1511.02
2	Toluene	0.012860	725.489	0.564·10^{-7}	3586.430
3	m-Xylene	0.005925	1023.69	236.674	−2255.3

For the extended Langmuir isotherm, the analytical form of the derivative is obtained as follows:

$$\frac{\partial y_i^*}{\partial X_i} = \frac{X_i(1 + \sum_{i=1, i \neq j} b_i \exp(c_i/T) \, P \cdot y_i^*)}{(X_{i0} \exp(a_i/T) - X_i)^2 b_i \exp(c_i/T) P}$$

$$+ \frac{1 + \sum_{i=1, i \neq j} b_i \exp(c_i/T) P \cdot y_i^*}{(X_{i0} \exp(a_i/T) - X_i) b_i \exp(c_i/T) P} \quad (15)$$

The appropriate boundary and initial conditions are as follows:

$$\frac{\partial y_i}{\partial z} = 0; \quad \text{for } z = 0 \quad (16a)$$

$$\frac{\partial P}{\partial z} = 0; \quad (16b)$$

$$T = T_0 = \text{const.} \quad (16c)$$

$$\frac{\partial T}{\partial z} = 0; \quad \text{for } z = L \quad (17a)$$

$$P = P(t) \quad (17b)$$

$$X_i = X_i(z, 0); \quad \text{for } t = 0 \quad (18a)$$

$$T = T(z, 0); \quad (18b)$$

$$P = P(z, 0). \quad (18c)$$

The above equations can be made dimensionless using the following normalized variables:

$$\tau = \frac{a_e t}{L^2}, \quad Z = \frac{z}{L}, \quad \omega = \frac{P - P_0}{\Delta P}, \quad \theta = \frac{T - T_0}{\Delta T},$$

$$\Omega_i = \frac{X_i - X_i^*}{X_{ip} - X_i^*}, \quad C_i^* = \frac{y_i C}{C_{i0}} \quad (19)$$

where the dependent variables have been normalized with respect to arbitrarily chosen reference variations (ΔP and ΔT).

The resulting dimensionless governing equations are written as:

$$a_e C_{i0} \frac{\partial C_i^*}{\partial \tau} + L \frac{\partial (y_i \cdot J_z)}{\partial Z} + \frac{a_e \rho_b (X_{ip} - X_i^*)}{\varepsilon M_i} \frac{\partial \Omega_i}{\partial \tau} = 0 \quad (20)$$

$$\frac{\partial \Omega_i}{\partial \tau} = -\frac{L^2}{a_e} K_i \Omega_i \quad (21)$$

$$-\frac{k_e}{a_e} \frac{\partial^2 \theta}{\partial Z^2} + c_\Sigma \frac{\partial \theta}{\partial \tau} + \frac{(X_{ip} - X_i^*) \rho_b}{\Delta T} \left(\sum_{i=1}^{s} \frac{\Delta H_i}{M_i} \right) \frac{\partial \Omega_i}{\partial \tau} = 0 \quad (22)$$

$$\Delta P \frac{\partial (\omega)}{\partial Z} = -\frac{L \mu}{C k_D} J_z - \left(\sum_{i=1}^{s} y_i M_i \right) \frac{L k_E}{C k_D} J_z^2 \quad (23)$$

where the dimensionless gas pressure has been used as an unknown dependent variable.

3 Solution of the problem

By knowing the equilibrium moisture content of components in the material, the proposed model was solved by the numerical method of lines, which is a general technique for solving partial differential equations by typically using finite difference relationships for the spatial derivatives and ordinary differential equations for the time derivative, as discussed by Schiesser (1991).

For a numerical problem solution, the drying material bed of thickness L is divided (Fig. 1) into N sections with $N + 1$ node points.

For this problem with layer thickness of $L = 0.005, 0.010, 0.015$ m, the number of divided sections are $N = 5, 10, 15$ with $\Delta z = 0.001$ m. Using a backward difference formula for the first and second derivatives, the mass balance equation for component "i" can be written as:

$$\frac{d (C_i^*)_j}{d\tau} = -\frac{L}{a_e C_{i0}} \frac{(y_i \cdot J_z)_{j+1} - (y_i \cdot J_z)_{j-1}}{2\Delta Z}$$
$$- \frac{\rho_b (X_{ip} - (X_i^*)_j)}{C_{i0} \varepsilon M_i} \frac{d(\Omega_i)_j}{d\tau} \quad \text{for } i = 1, \ldots, s; \quad j = 2, \ldots, N-1 \quad (24)$$

The mass balance equation for porous materials to be dried:

$$\frac{(d\Omega_i)_j}{d\tau} = -\frac{L^2}{a_e} K_i (\Omega_i)_j \quad \text{for } i = 1, \ldots, s; \quad j = 2, \ldots, N-1 \quad (25)$$

The heat balance equation:

$$\frac{d\theta_j}{d\tau} = \frac{k_e}{c_\Sigma a_e} \frac{\theta_{j+1} - 2\theta_j + \theta_{j-1}}{\Delta Z^2}$$
$$- \frac{(X_{ip} - X_i^*) \rho_b}{c_\Sigma \Delta T} \left(\sum_{i=1}^{s} \frac{\Delta H_i}{M_i} \right) \frac{d(\Omega_i)_j}{d\tau} \quad \text{for } i=1,\ldots,s; \quad j=2,\ldots,N-1 \quad (26)$$

Fig. 2 Average moisture content versus time. System: water (1) – zeolite DAY 20F. $T_0 = 373$ K; $L = 0.0100$ m; $P = 10$ Pa

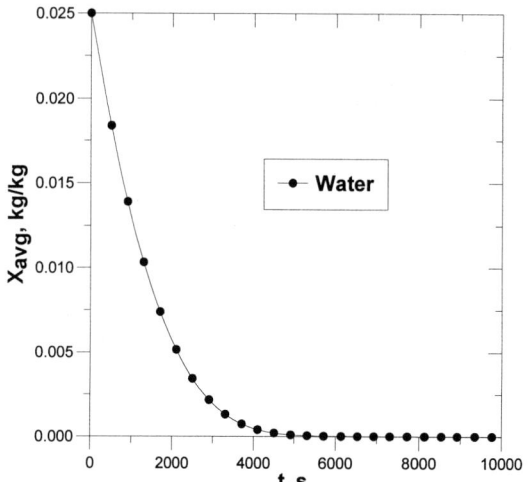

The pressure drop relation:

$$\frac{\omega_{j+1} - \omega_{j-1}}{2\Delta Z} = -\frac{L\,\mu}{\Delta P C k_D} J_z$$

$$-\left(\sum_{i=1}^{s} y_i M_i\right) \frac{L\,k_E}{\Delta P\,C\,k_D} J_z^2 \quad \text{for } i=1,\ldots,s;\ j=2,\ldots,N-1 \quad (27)$$

and the explicit relation for the molar vapor flux density is:

$$J_z = -\frac{2 C k_D \mu^{-1}\left[(\omega_{j+1} - \omega_{j-1})/2\Delta Z\right]}{1 + \sqrt{1 + 4C\left(\sum_{i=1}^{s} y_i M_i\right) k_D k_E \mu^{-2}\left|\left[(\omega_{j+1} - \omega_{j-1})/2\Delta Z\right]\right|}} \quad (28)$$

The difference scheme for method of lines is evident for elements $1 \leq j \leq n-1$. For $j = N$, the dimensionless pressure is equal to dimensionless pressure in the vacuum chamber $\omega_{ch} = (P_{ch} - P_0)/\Delta P = \text{const}$. But for element 0 a different difference scheme should be used. For $j = 0$ differential $\left.\frac{\partial \omega_j}{\partial Z}\right|_{j=0}$ is replaced by a forward difference instead of the central one:

$$\frac{\partial \omega_j}{\partial Z} = \frac{3\omega_j - 4\omega_{j+1} + \omega_{j+2}}{2\Delta Z} = 0 \quad (29)$$

which can be solved for ω_0 to yield $\omega_0 = (4\omega_1 - \omega_2)/3$.

Temperature and concentration at nodes 0 and N were computed using suitable boundary conditions, according to Eqs. (16a–c) and (17a, 17b).

As a result, concentration and temperature distributions of the mixture components versus time were computed.

The accuracy of the numerical solution was satisfactory. The solution was also obtained for double the number of sections and it was found that the concentration and temperature profiles were virtually unchanged.

As a result, concentrations in both phases of the mixture components and temperature distributions versus time were computed.

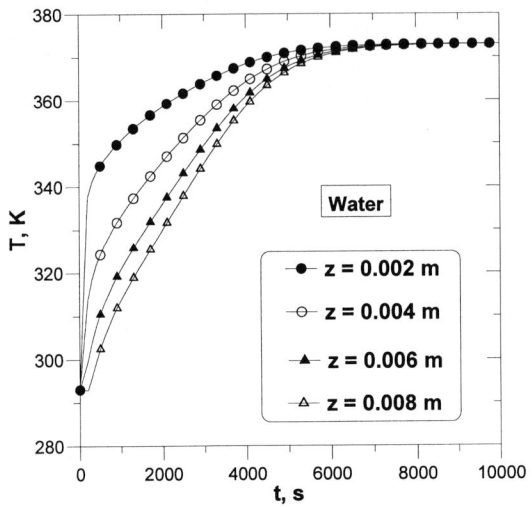

Fig. 3 Temperature versus time, for various positions in material. System: water−zeolite DAY 20F. $T_0 = 373$ K; $L = 0.0100$ m; $P = 10$ Pa

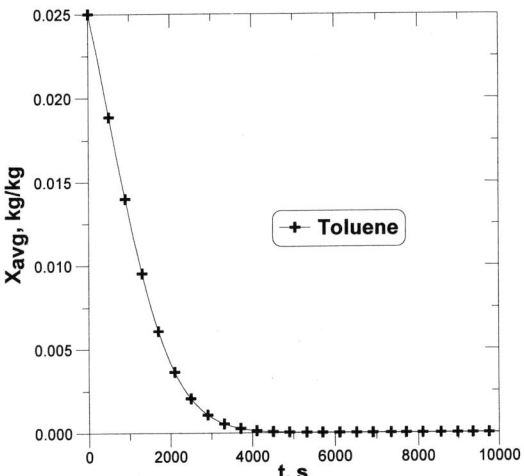

Fig. 4 Average moisture content versus time. System: toluene−zeolite DAY 20F. $T_0 = 373$ K; $L = 0.0100$ m; $P = 10$ Pa

Calculated characteristics of moisture content $X_i(z,t)$ and temperature $T(z,t)$ as a function of position and time were performed for one-, two-, and three-component mixtures: water (1)−toluene (2)−m-xylene (3) adsorbed on zeolite DAY 20F.

The simulations were done for the following process parameters: $L = 0.0025$; 0.0050, and 0.0100 m; $c_s = 900$ J/(kg K); $\rho_b = 500$ kg/m^3; $\varepsilon = 0.677$; $T_0 = 373$ K; $d_p = 0.5 \cdot 10^{-3}$ m; $T_i = 293$ K. In the vacuum desorption process the effective thermal conductivity k_e of the material to be dried depends strongly on total pressure P. The values of $k_e = 0.020$ W/(m K), $k_e = 0.025$ W/(m K) and $k_e = 0.030$ W/(m K) correspond to the pressures $P = 10$, $P = 100$, and $P = 1000$ Pa, respectively.

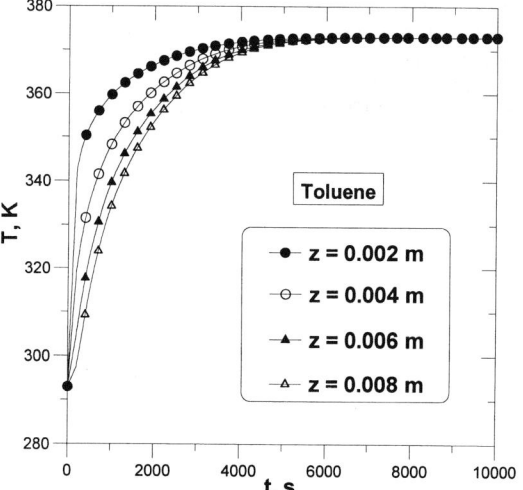

Fig. 5 Temperature versus time, for various positions in material. System: toluene—zeolite DAY 20F. $T_0 = 373$ K; $L = 0.0100$ m; $P = 10$ Pa

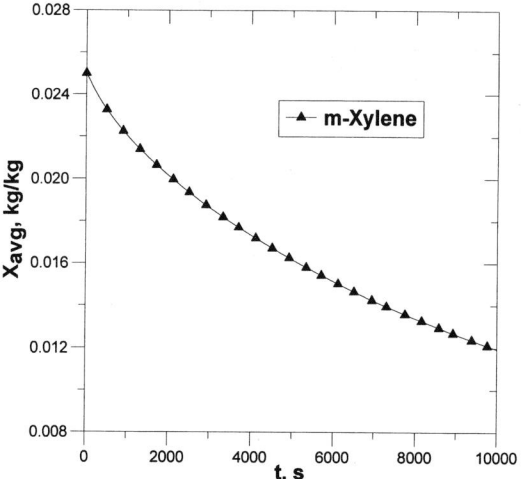

Fig. 6 Average moisture content versus time. System: m-xylene—zeolite DAY 20F. $T_0 = 373$ K; $L = 0.0100$ m; $P = 10$ Pa

4 Results and discussion

The simulation results of the vacuum desorption of pure water, toluene and m-xylene on zeolite DAY 20F are shown in Figs. 2–7.

The average moisture contents of the pure water, toluene, and m-xylene versus time during secondary vacuum freeze-drying at contact surface temperature $T_0 = 373$ K are shown in Figs. 2, 4, and 6, respectively.

Figures 3, 5, and 7 show the temperatures versus time for various positions in the material layer during desorption of pure species.

Analogous results for the vacuum desorption of two-component mixtures: water (1) toluene (2) water (1)—m-xylene (3), and toluene (2)—m-xylene (3) on zeolite DAY 20F are presented in Figs. 8–13.

Fig. 7 Temperature versus time, for various positions in material. System: m-xylene (3)—zeolite DAY 20F. $T_0 = 373$ K; $L = 0.0100$ m; $P = 10$ Pa

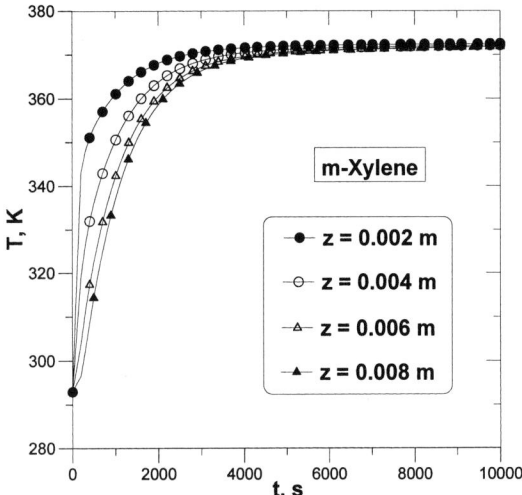

Fig. 8 Average moisture content versus time. System: water (1)—toluene (2)—zeolite DAY 20F. $T_0 = 373$ K; $L = 0.0100$ m; $P = 10$ Pa

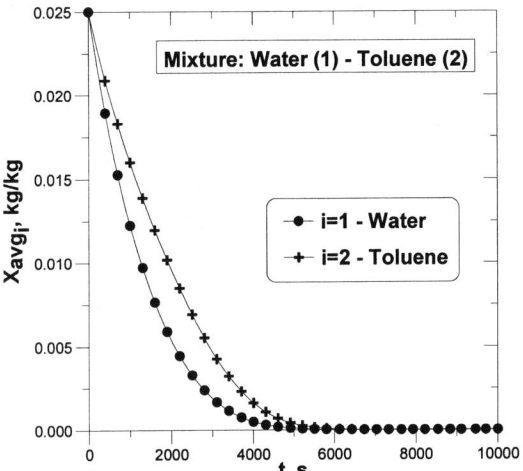

Figs. 8, 10, and 12 present the average moisture contents versus time for various positions in the material layer during two-component mixture desorption: water (1)—toluene (2), water (1) – m-xylene (3), and toluene (2)—m-xylene (3), respectively.

The temperatures versus time for various positions in the material layer during desorption of the above two-component mixtures are depicted in Figs. 9, 11, and 13, respectively.

Calculated results of the vacuum desorption of three-component mixture water (1)—toluene (2)—m-xylene (3) on zeolite DAY 20F are presented in Figs. 14–17.

The average moisture content of the water (1), toluene (2) and m-xylene (3) versus time during secondary vacuum freeze-drying at contact surface temperature $T_0 = 373$ K is shown in Fig. 14.

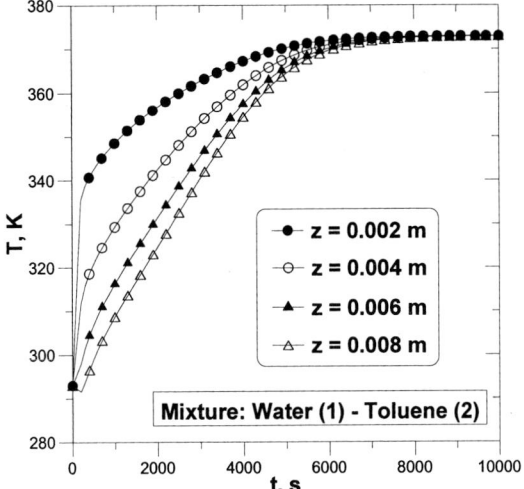

Fig. 9 Temperature versus time, for various positions in material. System: water (1)−toluene (2)−zeolite DAY 20F. $T_0 = 373$ K; $L = 0.0100$ m; $P = 10$ Pa

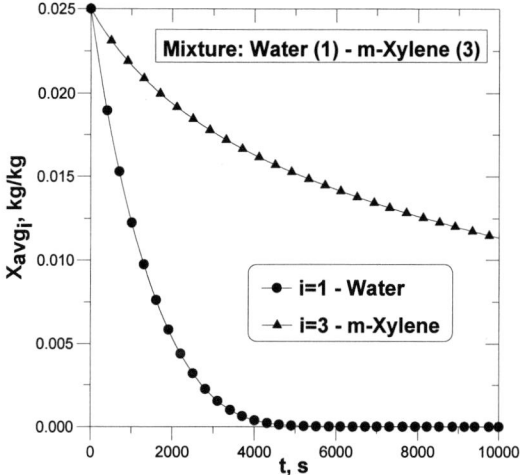

Fig. 10 Average moisture content versus time. System: water (1)−m-xylene (3)−zeolite DAY 20F. $T_0 = 373$ K; $L = 0.0100$ m; $P = 10$ Pa

Figure 15 presents the temperatures versus time for various positions in the material layer. A characteristic temperature decrease of the subsequent material layers can be observed here; this is due to consumption of the desorption heat "in situ" by removed components.

The influence of total pressure P in the vacuum chamber on the material temperature for axial coordinate $z = 0.0060$ m is depicted in Fig. 16.

Average water content versus time, for various material layer thicknesses $L = 0.0025$; 0.0050 and 0.0100 m with constant contact surface temperature $T_0 = 373$ K is given in Fig. 17.

An approach is proposed in the paper to obtain a mixed vapor isotherm from empirical single component vapor isotherms. The first step is approximation of the pure vapor isotherm by a multitemperature Langmuir correlation for single components water, toluene and m-xylene. The second step is application of an extended

Fig. 11 Temperature versus time, for various positions in material. System: water (1)—m-xylene (3)—zeolite DAY 20F. $T_0 = 373$ K; $L = 0.0100$ m; $P = 10$ Pa

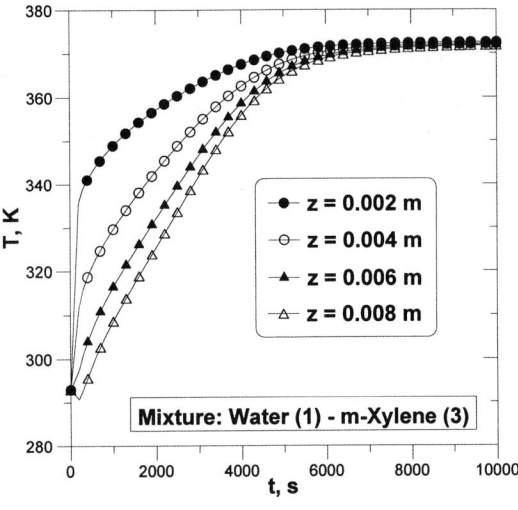

Fig. 12 Average moisture content versus time. System: toluene (2)—m-xylene (3)—zeolite DAY 20F. $T_0 = 373$ K; $L = 0.0100$ m; $P = 10$ Pa

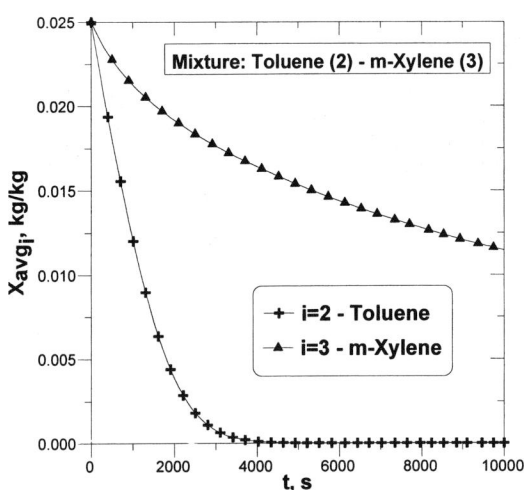

multitemperature Langmuir correlation to analytical prediction of the mixture isotherms, using only coefficients determined from single component Langmuir isotherms.

On the basis of the mixture isotherms one can choose parameters of the process (P, T_0), which enable control of the dried bed selectivity and secure desirable final composition.

The proposed analysis permits a theoretical prediction of the vacuum desorption process of mixture components contained in the material. The results should be verified by means of experimental investigations.

5 Conclusions

In this paper, a mathematical model of vacuum desorption of one-component and multicomponent mixtures adequate for secondary freeze-drying with conductive

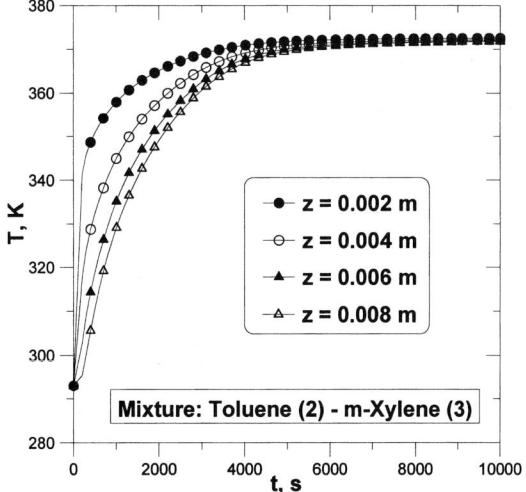

Fig. 13 Temperature versus time, for various positions in material. System: toluene (2)−m-xylene (3)−zeolite DAY 20F. $T_0 = 373$ K; $L = 0.0100$ m; $P = 10$ Pa

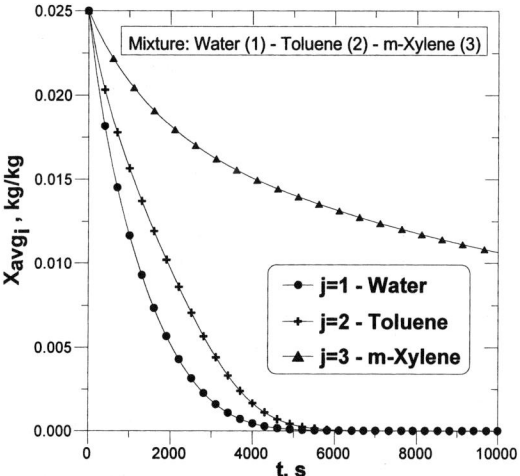

Fig. 14 Average moisture content versus time. System: water (1)−toluene (2)−m-xylene (3)−zeolite DAY 20F. $T_0 = 373$ K; $L = 0.0100$ m; $P = 10$ Pa

heating has been developed, taking especial account of pressure drop in the drying material layer.

The model also takes account of mass transfer resistances, which arise in vacuum desorption. The LDF mass transfer model with a variable, lumped-resistance coefficient K_i was used. This numerical model was used for computer simulation of the vacuum desorption of pure water, toluene, and m-xylene and their two- and three-component mixtures from zeolite DAY 20F.

The experimental determination of multicomponent adsorption equilibria is expensive and very time consuming, due to many possible combinations of concentrations of the adsorptives. It is possible to predict multicomponent adsorption equilibria using pure component equilibria data. Thus, the extended multitemperature Langmuir model has been used to predict adsorption for mixtures from single component isotherms.

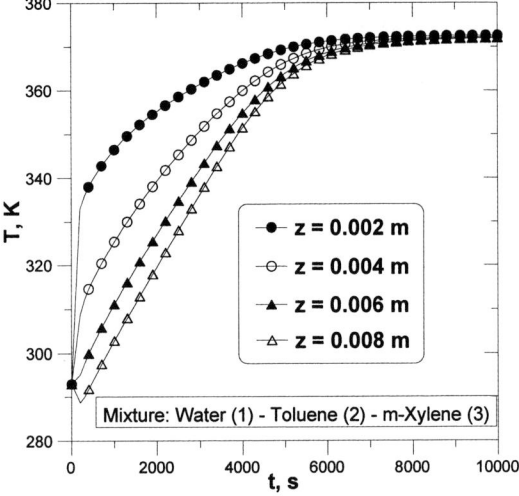

Fig. 15 Temperature versus time, for various positions in material. System: water (1)—toluene (2)—m-xylene (3)—zeolite DAY 20F. $T_0 = 373$ K; $L = 0.0100$ m; $P = 10$ Pa

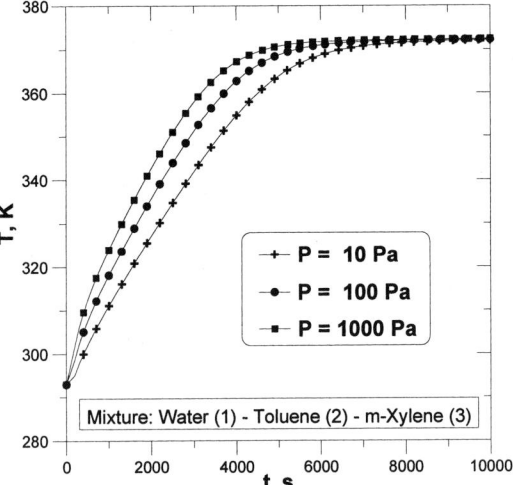

Fig. 16 Influence of total pressure P in the vacuum chamber on the material temperature in axial coordinate $z = 0,0060$ m. System: water (1)—toluene (2)—m-xylene (3)—zeolite DAY 20F. $T_0 = 373$ K; $L = 0.0100$ m

Theoretical analysis of the fixed bed vacuum desorption from the multicomponent moisture shows that the most important problem is the reliable prediction of the multitemperature mixed component equilibria on a specific kind of adsorbent.

In the literature, practically only rare two- and three-component mixture isotherms data exists. In addition, a lack of data for systems with water as a component is observed.

The method presented should be extended to quaternary and more complicated systems. This will require reliable empirical data for multitemperature pure component isotherms to predict mixture isotherms. Further theoretical and empirical investigations should be done to explain vacuum desorption from multicomponent moisture in close relation to real drying systems.

Fig. 17 Average water content versus time, for various dried layer thicknesses. System: water (1) − toluene (2) − m-xylene (3) − zeolite DAY 20F. $T_0 = 373\,\text{K}$; $P = 10\,\text{Pa}$

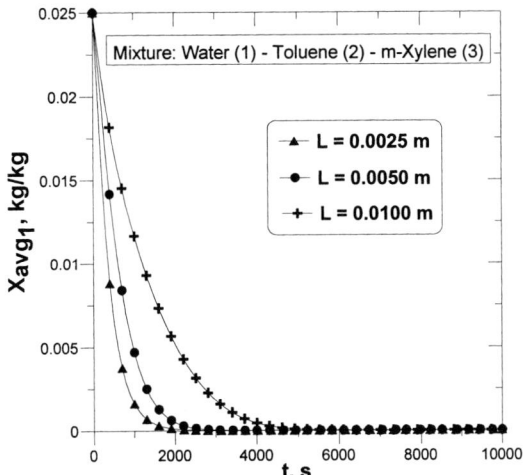

Acknowledgement The work was performed under financial support of grant No. 4 T09C 032 23 from Polish State Committee for Scientific Research.

References

Borger, T., Salden, A., Eigenberger, G.: A combined vacuum and temperature swing adsorption process for recovery of amine from foundry air. Chem. Eng. Proc. **36**, 231–244 (1997)
Brihi, T.A., Jaubert, J.-N., Barth, D.: Determining volatile organic compounds' adsorption isotherms on dealuminated Y zeolite and correlation with different models. J. Chem. Eng. Data **47**, 1553–1557 (2002)
Chahbani, M.H., Tondeur, V.: Pressure drop in fixed-bed adsorbers. Chem. Eng. J. **81**, 23–34 (2001)
Do, D.D.: Adsorption Analysis: Equilibria and Kinetics. Imperial College Press, London (1998)
Ergun, S.: Fluid flow through packed column. Chem. Eng. Prog. **48**, 89–94 (1952)
Glueckauf, E.: Theory of chromatography. Trans. Faraday Soc. **51**, 1540–1551 (1955)
Goldblith, S.A., Rey, L., Rothmayr, W.W.: Freeze-drying and Advanced Food Technology. Academic Press, London (1975)
Hall, K.R., Eagleton, L.C., Acrivos, A., Vermeulen, T.: Pore- and Solid-Diffusion Kinetics in Fixed bed Adsorption under Constant-Pattern Conditions. Academic Press, London (1966)
Hill, T.L.: Statistical mechanics of adsorption. V. thermodynamics and heat of adsorption. J. Chem. Phys. **17**, 520–535 (1949)
Kim, M.-B., Ryu, Y.-K., Lee, C.-H.: Adsorption equilibria of water vapor on activated carbon and DAY zeolite. J. Chem. Eng. Data **50**, 951–955 (2005)
Laurent, S., Couture, F., Roques, M.: Vacuum drying of a multicomponent pharmaceutical product having different pseudo-polymorphic forms. Chem. Eng. Proc. **38**, 157–165 (1999)
Liapis, A.I., Bruttini, R.: Freeze-drying of pharmaceutical crystalline and amorphous solutes in vials: Dynamic multi-dimensional models of the primary and secondary drying stages and qualitative features of the moving interface. Dry. Technol. **13**, 43–72 (1995)
Liapis, A.I., Pikal, M.J., Bruttini, R.: Research and developments needs and opportunities in freeze drying. Dry. Technol. **14**, 1265–1300 (1996)
Nastaj, J., Ambrożek B.: Modeling of vacuum desorption in freeze drying process. Dry. Technol. **23**, 1693–1709 (2005)
Poling, B.E., Prausnitz, J.M., O'Connell, J.P.: The Properties of Gases and Liquids. – 5th ed., McGraw-Hill, New York (2000)
Rowe, T.W., Snowman, J.W.: Edwards Freeze-Drying Handbook. Crawley, Cambridge (1978)
Ruthven, D.M.: Principles of Adsorption and Adsorption Processes. Wiley Interscience, New York (1984)

Ryu, Y.-K., Chang, J.-W., Jung, S.-Y., Lee, C.-H.: Adsorption isotherms of toluene and gasoline vapors on DAY zeolite. J. Chem. Eng. Data **47**, 363–366 (2002)

Sadikoglu, H., Lapis, A.I.: Mathematical modeling of the primary and secondary drying stages of bulk solution freeze-drying in trays: parameter estimation and model discrimination by comparison of theoretical results with experimental data. Dry. Technol. **15**, 791–810 (1997)

Schiesser, W.E.: The Numerical Method of Lines: Integration of Partial Differential Equations. Academic Press, New York (1991)

Sun, L.M., Ben Amar, N., Meunier, F.: Numerical study on coupled heat and mass transfer in an adsorber with external fluid heating. *Heat Recovery Systems & CHP* **15**, 19–29 (1995)

Wolf, H.E., Schlünder, E.U.: Adsorption equilibrium of solvent mixtures on silica gel and silica gel coated ceramics. Chem. Eng. Process **38**, 211–218 (1999)

Yao, C.: Extended and improved Langmuir equation for correlating adsorption equilibrium data. Sep. Purif. Technol. **19**, 237–242 (2000)

Transp Porous Med (2007) 66:219–231
DOI 10.1007/s11242-006-9016-0

ORIGINAL PAPER

Optimization of fine solid drying in bubble fluidized bed

Artur Poświata · Zbigniew Szwast

Received: 20 October 2005 / Accepted: 26 March 2006 /
Published online: 30 August 2006
© Springer Science+Business Media B.V. 2006

Abstract A commonly used method to dry fine solid particles is drying in a fluidized bed. This paper presents the optimization problem of fluidized drying of fine solids. A drying process proceeding in a three-stage cascade of fluidized cross-current dryers was considered. Solid flows from stage to stage, and fresh gas is introduced to each stage of the cascade. The hydrodynamics of bubble fluidized bed and kinetics of heat and mass transfer are taken into account. The bed hydrodynamics is described by a two-phase model. The drying process considered proceeds in the second period of drying. To optimize this problem a generalized version of a discrete algorithm with constant Hamiltonian was used. The optimization procedure is presented in the paper. In optimization calculations, gas parameters (temperature, humidity and flow rate) minimizing total process cost are sought. The results of calculation are presented as graphs. The results obtained and the conclusions drawn are discussed.

Keywords Fluidized drying · Drying optimization · Exergy minimizing · Hamiltonian · Optimization algorithm

Nomenclature
A, B Coefficients in performance index, kJ/(kg K^2),kJ/kg
c_g Gas specific heat , kJ/(kg K)
G Gas mass flow rate, kg/s
h Solid enthalpy, kJ/kg
h_{mf} Bed height in minimum fluidization, m
i Gas enthalpy, kJ/kg
i_{vb} Moisture enthalpy in bubble phase, kJ/kg
i_{vmf} Moisture enthalpy in dense phase, kJ/kg

A. Poświata (✉) · Z. Szwast
Department of Chemical and Process Engineering,
Warsaw University of Technology,
Warsaw, Poland
e-mail: poswiata@ichip.pw.edu.pl

K	Drying constant, 1/s
k_b	Mass transfer coefficient between dense and bubble phase, kg/m^3
S	Solid mass flow rate, kg/s
S_v	Volume solid flow rate, m^3/s
T	Temperature, °C, K
u_g	Gas superficial velocity, m/s
u_{mf}	Minimum fluidization velocity, m/s
u_p	Velocity of gas in bubble phase, m/s
W_b	Dimensionless resistance coefficients of mass transfer
X_e	Equilibrium moisture content in solid
X_s	Moisture content in solid
Y	Moisture content in gas
Z_b	Dimensionless resistance coefficients of heat transfer between gas and solid
Z_{mf}	Dimensionless resistance coefficients of heat transfer between bubble and dense phases

Greek symbols

α_b	Heat transfer coefficient between dense and bubble phase, kJ/m^{3}K
α_{mf}	Heat transfer coefficient between solid and gas, kJ/m^{3}K
κ	Exergy coefficient of investment and gas pumping costs, kJ/kg
ρ	Density, kg/m^3
ζ	Ratio of chemical exergy unit price to thermal exergy unit price
τ	Average solid hold-up, s
σ	Volume fraction of bubble phase in fluidized bed
θ_g	Dimensionless gas flow rate (ratio of gas mass flow rate to solid one)
σ_g	Mass fraction of gas flowing through bubble
ε_{mf}	Dense phase porosity

Subscripts

b	Outlet gas from bubble phase
g	Inlet drying gas
k	Outlet drying gas
mf	Gas in dense phase
s	Solid

Superscripts

n	Stage number

1 Introduction

Drying, as one of the most expensive and energy-consuming industrial processes, can strongly influence total production costs. Simultaneously, as one of the terminal processes, the drying process influences the quality of the final product, so reduction of drying costs without product deterioration is a significant challenge to engineers and process designers. Taking into account the high costs of drying, reduction of these costs by even a few percent can bring considerable economic gains.

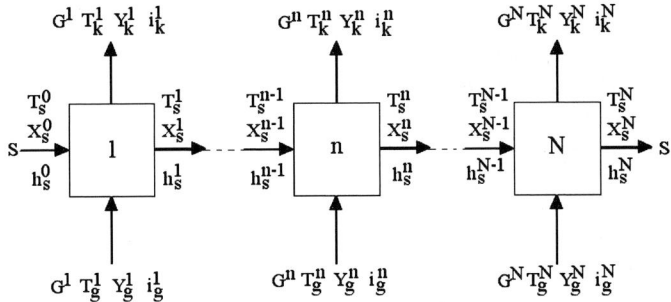

Fig. 1 Scheme of multistage fluidized drying of fine solids

One of the ways to reduce a process cost is optimization, i.e. a choice of such process parameters as minimize the cost of the process. In cost optimization, both investment costs and exploitation costs must be taken into consideration. A mathematical description of these costs requires a knowledge of many economic data. Unfortunately, economic data are often fragmentary, and existing data are dubious with respect to local and time fluctuations. Investigations are therefore undertaken of the use of thermodynamic criteria of economic values in cost optimization. Based on thermodynamic and economic considerations, one accepts that the economic value of a medium used in a process is proportional to the exergy of this medium, so economic optimization criteria can be replaced by exergy criteria, i.e., a function describing the economic cost of a process can be substituted by an equivalent thermodynamic function (exergy). This approach enables one to suppress the influence of local and temporal fluctuations of economic data and to generalize some optimization results. Hence, in optimization calculations the performance index describes a total cost of the process expressed in exergy units. Classical drying process indicators, such as energy utilization coefficient, heat consumption per kg of water evaporation or heat consumption per kg of dry product, are not used in this paper because these indicators do not take account of investment cost, utilization time and other economic quantities.

2 Optimization problem definition

This paper considers an optimization problem of multistage fluidized drying of fine solid in the second period of drying. Fresh drying gas is provided to all stages, whereas solid flows from stage to stage (Fig. 1). In optimization calculations, optimum values of drying gas parameters (temperature, humidity and flow rate) for all stages must be found. Optimum values of gas parameters minimize a performance index describing the total cost of the process. Moreover, values of solid temperatures and humidities between stages must be found, too. For typical drying processes, initial solid temperature and moisture content, as well as final solid moisture content are known. The final solid temperature is free, so the temperature must be calculated during optimization.

The performance index accepted for calculation describes the total costs of drying per unit of dry solid mass flow. The process cost is expressed in exergy units. The performance index derived by Sieniutycz and Szwast (Berry et al. 2000) and modified by the present authors takes the form of Eq. 1, in which the first term on the right-hand side describes the thermal exergy of drying gas connected with the temperature of this gas (higher than ambient), the second term describes a chemical exergy connected with humidity of drying gas (lower than ambient) and the third term, κ, describes investment cost and gas pumping cost:

$$I = \sum_{n=1}^{N} \left[\frac{1}{2} A \left(T_g^n - T_a \right)^2 + \frac{1}{2} \zeta B \left(Y_g^n - Y_a \right)^2 + \kappa \right] \theta_g^n \qquad (1)$$

where N is a stage number; A and B are coefficients Berry et al. (2000); T_g^n and Y_g^n are drying gas inletting to stage n temperature and humidity, respectively; T_a and Y_a are ambient temperature and humidity, respectively; ζ is a ratio of unit price of drying gas chemical exergy to unit price of drying gas thermal exergy; κ is a so-called exergy coefficient of investment and gas pumping costs; and θ_g^n is a dimensionless gas flow rate on stage n, defined as a ratio of gas mass flow rate to solid flow rate.

3 Fluidized drying model

The hydrodynamics of a fluidized bed is described by a Kunii and Levenspiel two-phase fluidized bed model (Kunii and Levenspiel 1991). The following assumptions are made in the model: the fluidized bed consists of two phases, a dense phase and a bubble phase. The dense phase consists of solid and gas: both are ideally mixed. In the dense phase, the conditions of fluidization are the same as in the minimum fluidization state (e.g. porosity, gas and velocity). Gas flowing by the dense phase has a minimum fluidization velocity. The bubble phase is formed by bubbles of pure gas flowing up through the dense phase. Plug flow of gas is assumed in the bubble phase the. The gas bubbles have a constant diameter along the bed height. The last assumption can be accepted because the fluidization bed height in drying processes is relatively low.

The volume fraction of bubble phase in a fluidized bed, σ, and mass fraction of gas flowing by the bubble phase are described by the following forms

$$\sigma = \frac{u_g - u_{mf}}{u_p - u_{mf}}, \qquad (2)$$

$$\sigma_g = 1 - u_{mf} \frac{1 - \sigma}{u_g}, \qquad (3)$$

where u_{mf}, u_g and u_p are minimum fluidization velocity, superficial gas velocity and gas velocity in the bubble phase, respectively. One can find many literature correlations to calculate the value of u_{mf}, whereas the method to determine u_p is presented in Kunii and Levenspiel's handbook (Kunii and Levenspiel 1991).

The second period of drying is considered in this paper. For the second period of drying the main resistances of moisture transfer are inside solids and any mass transfer resistances in the gas are neglected. So gas humidity on the interfacial surface (surface of solid particles) is the same as deep inside the gas phase. The driving force of moisture transfer is expressed as the difference between current solid moisture content and equilibrium solid moisture content, being a function of gas parameters on the interfacial surface. Moreover, the following assumption is made for heat transfer: heat conduction inside solid particles is very fast, so only the resistance of heat transfer in the gaseous phase is taken into account.

The drying rate is described by following form for $n = 1 \ldots N$

$$\frac{dX_s^n}{dt} = -K^n \left(X_s^n - X_e^n \right), \tag{4}$$

where K^n is a drying constant; X_s^n and X_e^n are current humidity of solid and equilibrium solid moisture content, respectively.

If the drying process were a batch process with initial solid moisture content equal to the inlet moisture content of the solid for stage n, X_s^{n-1}, and final moisture content of solid equal to the outlet moisture content of the solid for stage n, X_s^n, the following equation would be obtained after integration of Eq. 4 from X_s^{n-1} to X_s^n and from 0 to final time, τ^n:

$$\frac{X_s^{n-1} - X_e^n}{X_s^n - X_e^n} = e^{K^n \tau^n}. \tag{5}$$

Although the process considered in this paper is a flow, for the individual particle of solid it is a batch process as each particle passes through every state between X_s^{n-1} and X_s^n. We assume that the drying time of each particle for the given stage of cascade is the same for every solid particle and this time equals an average time of solid hold-up on stage n, τ^n. The solid hold-up time is expressed as the ratio of solid volume, V_s^n, to volume flow rate of solid, S_v:

$$\tau^n = \frac{V_s^n}{S_v}. \tag{6}$$

The time τ^n defined by Eq. 6 can be also expressed by some fluidized bed parameters and dimensionless gas flow rate. In effect, the average time of solid hold-up is a linear function of dimensionless gas flow rate, θ_g^n, and can be written in the following form:

$$\tau^n = \frac{h_{mf} \rho_s (1 - \varepsilon_{mf})}{\rho_g u_g} \theta_g^n = D_t \theta_g^n. \tag{7}$$

In investigations of the drying of black tea, Temple (Temple and van Boxtel 1999a, b; Temple et al. 2000) defined and used the following empirical equation describing the drying constant

$$K^n = a_1 \left(T_g^n - a_2 \right) + a_3, \tag{8}$$

where a_1, a_2 and a_3 are some empirical constants. Temple and van Boxtel (1999a) and Temple et al. (2000) proposed the following values for these constants if the temperature T_g^n is expressed in Celsius degrees and the gas velocity, u_g, in mps: $a_1 = 0.00028\, u_g$, $a_2 = 45$ and $a_3 = 0.00067$.

If time τ^n in Eq. (5) is expressed by Eq. (7) the following equation describing the change of solid moisture content on stage n is obtained:

$$X_s^n - X_s^{n-1} = \left(X_s^n - X_e^n \right) \left(1 - e^{K^n D_t \theta_g^n} \right). \tag{9}$$

Equations of mass and enthalpy balances for each stage of cascade are taken in the following forms (Poświata and Szwast 2003, 2004):

$$X_s^n - X_s^{n-1} = \theta_g^n \left[\sigma_g \left(Y_g^n - Y_b^n \right) + (1 - \sigma_g) \left(Y_g^n - Y_{mf}^n \right) \right], \tag{10}$$

$$h_s^n - h_s^{n-1} = \theta_g^n \left[\sigma_g \left(i_g^n - i_b^n\right) + (1 - \sigma_g) \left(i_g^n - i_{mf}^n\right)\right]. \tag{11}$$

The right-hand sides of Eqs. 10 and 11 follow from the fact that gas exhausting the fluidized bed flows through two phases. The first terms in square brackets describe changes of parameters of gas flowing through the bubble phase, whereas the second terms describe changes of parameters of gas flowing through dense phase.

Kinetic equations describing heat transfer between the gas in the dense phase and the solid takes the following form:

$$h_s^n - h_i^n = \frac{c_g \left(T_{mf}^n - T_s^n\right)}{Z_{mf}} \theta_g^n + \left(X_s^n - X_s^{n-1}\right) i_{vmf}^n. \tag{12}$$

The left-hand side of the above equation describes changes of solid enthalpy, the first term on the right-hand side describes heat transferred from gas to solid by convection, whereas the second one describes enthalpy transported from solid to gas together with a moisture stream. Coefficient Z_{mf} is expressed in the following form:

$$Z_{mf} = \frac{u_{mf} v_g (1 - \sigma) \rho_g c_g}{\alpha_{mf} h_{mf}}. \tag{13}$$

The Z_{mf} is a dimensionless resistance coefficient of heat transfer between gas and solid. The coefficient Z_{mf} can take positive values or zero. For $Z_{mf} = 0$ the heat resistance is equal to zero so heat transfer is very rapid and solid temperature is equal to the temperature of gas in the dense phase.

Mass transfer between the dense phase and the bubble phase is described by the the following equation:

$$Y_b^n - Y_g^n = -\frac{\Delta Y_b^n}{W_b}, \tag{14}$$

where W_b is a resistance coefficient of mass transfer between dense phase and bubble phase, whereas ΔY_b^n is a driving force for mass transfer expressed as a logarithmic average of differences of humidity between bubble phase and dense phase. The following form describes coefficient W_b:

$$W_b = \frac{u_g \rho_g \sigma_g (1 - \sigma)}{k_b h_{mf} \sigma}. \tag{15}$$

Enthalpy changes of gas flowing through bubble phase are expressed by Eq. 16:

$$i_g^n - i_b^n = \frac{c_g \Delta T_b^n}{Z_b} - \frac{\Delta Y_b^n}{W_b} i_{vb}^n, \tag{16}$$

where Z_b is dimensionless coefficient of heat transfer resistance between dense phase and bubble phase expressed by Eq. (17)

$$Z_b = \frac{c_g u_g \rho_g \sigma_g (1 - \sigma)}{\alpha_b h_{mf} \sigma} \tag{17}$$

whereas ΔT_b^n is a logarithmic average of differences of temperature between bubble phase and dense phase, i.e. the driving force for heat transfer between bubble and dense phases.

Coefficients Z_b and W_b take positive values or zero, similar to coefficient Z_{mf}. If these both coefficients equal 0 then thermodynamic equilibrium is achieved between gas in the bubble phase and gas in the dense phase. Moreover, equality of these coefficients ($Z_b = W_b$) can be accepted for the water–air systems commonly occurring in the drying processes.

Equations 9–12, 14 and 16 are the main equations of the mathematical model of the fluidized drying process of fine solid in the second period of drying. This set of equations does not have an analytical solution, so it must be solved numerically.

4 Optimization algorithm

The discrete algorithm with constant Hamiltonian known in the literature requires a mathematical model of the process to be linear with respect to one selected decision variable Berry et al. (2000) The model of the drying process considered does not satisfy this requirement, so optimization calculations were performed using a generalized version of the discrete algorithm with constant Hamiltonian (Poświata and Szwast 2001, 2004).

The fact that there are no analytical solutions for the mathematical model of the drying process considered means that analytical forms of state transformation (i.e. a function describing changes of state variables on stage n) cannot be found, so values of the state transformations and their derivatives must be calculated numerically. The state transformations for solid moisture content and temperature can be presented in the following general forms:

$$X_s^n - X_s^{n-1} = f_X^n \left(X_s^n, T_s^n, T_g^n, Y_g^n, \theta_g^n \right) \theta_g^n, \tag{18}$$

$$T_s^n - T_s^{n-1} = f_T^n \left(X_s^n, T_s^n, T_g^n, Y_g^n, \theta_g^n \right) \theta_g^n, \tag{19}$$

Equations 18 and 19 describe changes of solid moisture content and temperature on the stage n as a function of the state variables, X_s^n and T_s^n, and decision variables, T_g^n, Y_g^n and θ_g^n, on this stage. It should be noticed that decision variable θ_g^n is an argument of functions f, which is an extension of the algorithm with constant Hamiltonian known the literature (Berry et al. 2000).

Moreover, an additional state variable t^n is introduced. In the theory of optimization this variable is called a time, but one ought to stress that it need not be a physical time. In our case the t^n variable describes total gas flow rate from first stage of cascade to stage n. So state transformation for this variable takes the form:

$$t^n - t^{n-1} = \theta_g^n. \tag{20}$$

The Hamiltonian is defined as follow

$$H^{n-1} = \frac{1}{2} A \left(T_g^n - T_a \right)^2 + \frac{1}{2} \varsigma B \left(Y_g^n - Y_a \right)^2 + \kappa$$
$$+ z_X^{n-1} f_X^n + z_T^{n-1} f_T^n + z_t^{n-1}, \tag{21}$$

where z_X^{n-1} and z_T^{n-1} are so-called adjoint variables.

For an optimum value of the decision variables (gas temperature, humidity and flow rate), the following conditions have to be satisfied:

$$\frac{\partial H^{n-1}}{\partial T_g^n} = 0, \tag{22}$$

$$\frac{\partial H^{n-1}}{\partial Y_g^n} = 0, \tag{23}$$

$$\frac{\partial H^{n-1}}{\partial \theta_g^n} + \frac{H^{n-1}}{\theta_g^n} = 0. \tag{24}$$

Moreover, for an optimum process the following adjoint equations describing changes of adjoint variables have to be satisfied:

$$\frac{z_X^n - z_X^{n-1}}{\theta_g^n} = -\frac{\partial H^{n-1}}{\partial X_s^n}, \tag{25}$$

$$\frac{z_T^n - z_T^{n-1}}{\theta_g^n} = -\frac{\partial H^{n-1}}{\partial T_s^n}, \tag{26}$$

$$\frac{z_t^n - z_t^{n-1}}{\theta_g^n} = -\frac{\partial H^{n-1}}{\partial t^n}. \tag{27}$$

Thus, the set of equations, which has to be solved during the optimization procedure consists of the state transformations (18)–(20) and Eqs. 22 to 27.

To resolve the problem, we must have defined boundary conditions for state and adjoint variables. In typical drying problems, the initial values of state variables (solid temperature and humidity) are known, as well as final solid moisture content. The final solid temperature can be fixed or free, and total gas flow rate, t^N, can be fixed or free. This paper considers only cases with free final solid temperature and free total gas flow rate.

Boundary conditions for adjoint variables are as follows: if the final value of a state variable is fixed then the final value of the right adjoint variable is free (z^N is free), and if the final value of a state variable is free then the final value of the adjoint variable is zero ($z^N = 0$). Therefore, for fixed final solid moisture content the final value of the variable z_X^N is undetermined and for free final solid temperature variable z_T^N equals zero and also z_t^N equals zero for free total gas flow rate. Furthermore, the Hamiltonian considered is not an explicit function of variable t^n (autonomous process), so due to Equation 27 the adjoint variable z_t^n equals zero for all stages of the cascade.

Optimization calculations are performed in the reverse direction than the process run, i.e. from the last stage of the cascade to the first. At the beginning of calculation unknown values of state and adjoint variables for the end of the cascade (with superscript N) are assumed. The equations for the last stage, Eqs. 18–20 and 22–27 with superscript N, are then resolved. In effect, the values of the optimum gas parameters for the last stage and values of adjoint and state variables before the last stage (superscript $N-1$) are obtained, so the calculation for the penultimate stage can be performed. Continuing the calculation from stage to stage, values of state and adjoint

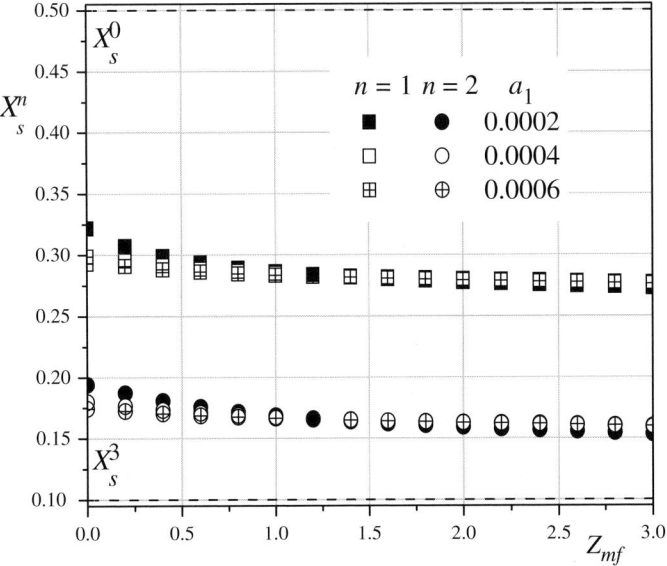

Fig. 2 Solid moisture content at all stages of cascade for optimum processes as a function of coefficient Z_{mf} (Dashed lines present an initial, X_s^0, and final, X_s^3, solid moisture content)

variables before the first stage are calculated. If the calculated values of the initial state are equal to fixed values of this state, the calculations are finished: if not, new values of state and adjoint variables after the last stage of cascade must be assumed.

5 Results of optimization calculations

The process of fluidized drying of fine solid in second period of drying is considered. The process proceeds in a three-stage cascade of a fluidized dryer. The following assumptions are made for initial and final values of process variables: the solid is dried from initial humidity $X_s^0 = 0.5$ kg of moisture per kg of dry matter to final humidity $X_s^3 = 0.1$ kg of moisture per kg of dry matter; the initial solid temperature is equal to ambient temperature $T_s^0 = T_a = 17°C$ whereas the final solid temperature, T_s^3, is free. The calculation was performed for exergy coefficient of investment and gas pumping costs $\kappa = 3$ kJ/kg and parameter $\zeta = 3$.

Figure 2 shows solid moisture content at all stages of the cascade for optimum processes as a function of the coefficient of heat transfer resistance between gas and solid, Z_{mf}, for three values of coefficient a_1. Drops of solid moisture content are the highest on the first stage and the least on the last (third) stage, but values of humidity decrease are of the same order of magnitude. Therefore, the use of multistage processes is well-founded if drying processes proceed in the second period of drying. The other situation is observed for drying of fine solid in the first period of drying, when optimum processes are single-stage (Poświata and Szwast 2001, 2003).

Moreover, analyzing Fig. 2, we can conclude that interstage solid moisture content is almost independent of values of coefficients Z_{mf} and a_1, so it is independent of heat and mass transfer kinetics and fluidized bed hydrodynamics.

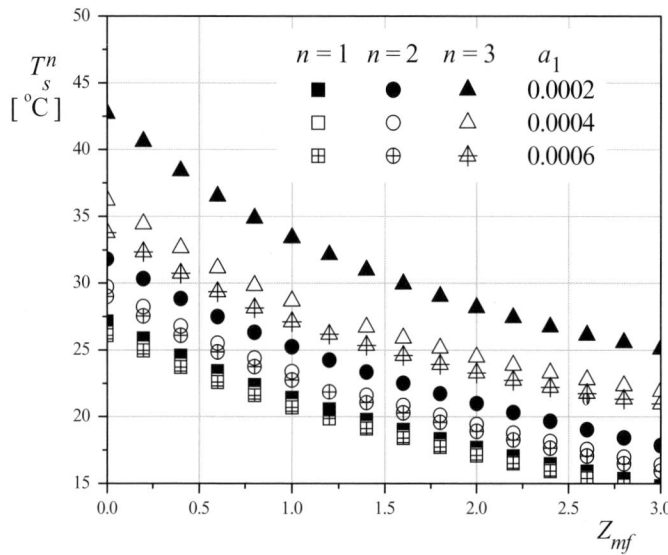

Fig. 3 Solid temperature on all stages for optimum process as a function of coefficient Z_{mf}

Figure 3 shows solid temperatures at every cascade stage for optimum processes as a function of coefficient Z_{mf} for three values of coefficient a_1. The values of these coefficients strongly influence solid temperatures. If heat transfer resistance increases (Z_{mf} increases) then solid temperature decreases for every stage. The same applies if coefficient a_1 increases (mass transfer resistances inside solid particles decreases), then solid temperature decreases. The solid temperature increases along the cascade.

In Figs 4 and 5 optimum values of gas temperature and humidity are presented as a function of resistance coefficient of heat transfer between solid and gas, Z_{mf}. Optimum values are shown for three assumed values of coefficient a_1. A converse of coefficient a_1 can be considered as a measure of mass transfer resistance inside solid particles. If a_1 takes high-values then the drying constant is high and mass transfer rapid, whereas if a_1 takes low values then mass transfer inside solid particles is slow. For fixed value of a_1 the optimum gas temperature and humidity decrease if heat transfer resistance increases (value of Z_{mf} increases). On the other hand, for fixed values of Z_{mf}, a decrease of a_1 (i.e. increase of mass transfer resistance) causes optimum gas temperature and humidity to increase.

Optimum drying gas temperature always increases as stage number increases, so the lowest temperature is on the first stage and the highest on the last one, Fig. 4. The changes of optimum gas humidity along the cascade are more complicated, Fig. 5. The course of optimum gas humidity along the cascade depends on values of coefficients Z_{mf} and a_1.

Optimum gas flow rate as a function of coefficient Z_{mf} is shown in Fig. 6 for three assumed values of coefficient a_1. Increase of heat transfer resistance, Z_{mf}, always causes an increase of gas flow rate, whereas in practice the optimum value of gas flow rate is independent of values of a_1. Moreover, we can state that dependence of optimum gas flow rate on coefficient Z_{mf} is linear for every stage of the cascade.

The maximum optimum gas flow rate is on the first stage of the cascade, decreases on, the next stage and so, on the last stage, the flow rate is a minimum.

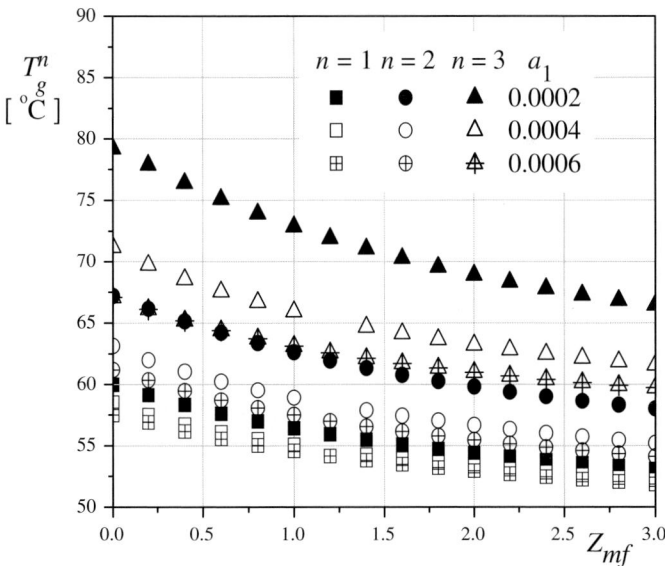

Fig. 4 Optimum gas temperatures as a function of coefficients Z_{mf} for three-stage cascade of fluidized drier

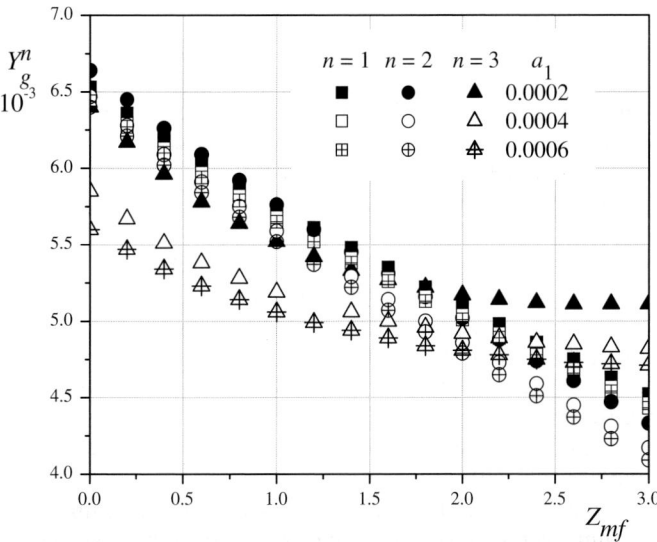

Fig. 5 Optimum gas humidities as a function of coefficients Z_{mf} for three-stage cascade of fluidized drier

6 Conclusions

The interstage solid humidities for optimum processes are practically independent of fluidized bed hydrodynamics and kinetics of heat and mass transfer. However, the

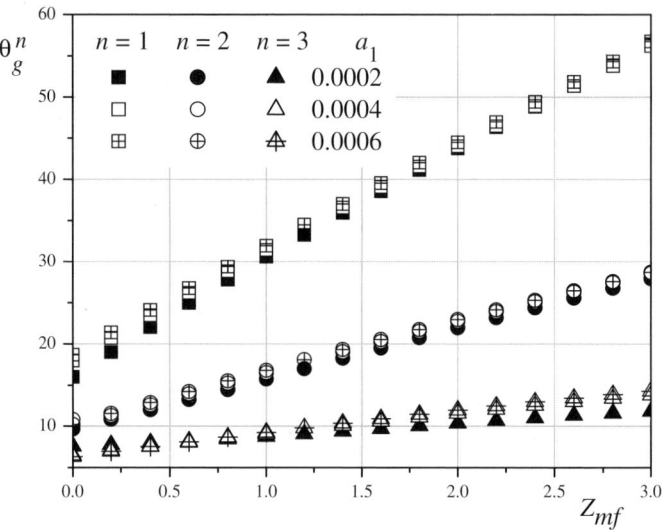

Fig. 6 Optimum gas flow rate as a function of coefficients Z_{mf} for three-stage cascade of fluidized drier

interstage solid temperatures are clearly dependent on the hydrodynamic and the kinetics.

Optimum values of temperature and humidity of drying gas decrease, respectively, for each stage of cascade if the value of the coefficient Z_{mf} increases as well as if the value of the coefficient a_1 increases. Optimum temperature of inlet drying gas increases with increment of stage number of the cascade (independent of values of Z_{mf} and a_1). Recalling that, for the first period of drying, the opposite situation is observed, i.e. the optimum temperature decreases as stage number increases (Poświata and Szwast 2001).

Variation of optimum gas humidity along the cascade is dependent on values of coefficients Z_{mf} and a_1 and is more complicated than dependence on gas temperature. For low values of Z_{mf} and a_1 optimum gas humidity is lowest on the third stage of the cascade and is highest on the second stage, whereas for high values of Z_{mf} and a_1 the situation is reversed i.e. optimum gas humidity is lowest on the second stage and highest on the third one.

For every stage of the cascade, optimum gas flow rate increases with the increment of value of coefficient Z_{mf}, whereas variation of optimum gas flow rate with the increment of coefficient a_1 has a different character for different stages of the cascade: for the first and the second stages, optimum gas flow rate increases with increasing coefficient a_1. For the third stage optimum gas flow rate decreases with increase of coefficient a_1 for low values of Z_{mf}, whereas it increases with increase of coefficient a_1 for high values of Z_{mf}. However, the dependence of optimum gas flow rate on coefficient a_1 is only weak, therefore in practice one can accept that optimum gas flow rate through all stages is independent of the value of coefficient a_1.

Optimum gas flow rate decreases with the increment of stage number in the cascade, but the optimum gas flow rates through every stages have comparable values. This is the opposite situation to that observed for fluidized drying of solid in the first period of drying, considered in papers (Poświata and Szwast 2001, 2003, 2004), in which almost all gas flow rate is connected with the first stage of the cascade.

The generalized version of the discrete algorithm with constant Hamiltonian presented here is a powerful method for optimizing any multistage process characterized by a non-linear mathematical model.

References

Berry R.S., et. al.: Thermodynamic optimization of Finite—Time Processes. Wiley, Chichester (2000)
Kunii D., Levenspiel O.: Fluidization Engineering. Butterworth, Heinemann, Newton, London (1991)
Poświata A., Szwast Z.: Optimization of heat and mass transfer in fluidized bed with bubble phase (in Polish). Inżynieria Chemiczna i Procesowa **22**(3D), 1187–1192 (2001)
Poświata A., Szwast Z.: Minimization of exergy consumption in fluidized drying processes. In: Proceedings of The 16th International Conference on Efficiency, Cost, Optimization, Simulation, and Enviromental Impact of Energy Systems, ECOS 2003, pp. 785–792. Copenhagen, Denmark, (2003)
Poświata A., Szwast Z.: Optimization of drying of solid being in second period of drying in bubble fluidized bed (in Polish). Inżynieria Chemiczna i Procesowa **25**, 151551–151556 (2004)
Temple J.S., van Boxtel A.J.B.: Equilibrium moisture content Tea. J. Agri. Eng. Res. **74**(1), 83–89 (1999a)
Temple J.S., van Boxtel A.J.B.: Thin layer drying of balck tea. J. Agri. Eng. Res. **74**(2), 167–176 (1999b)
Temple J.S., et al.: Monitoring and control of fluid—bed drying of tea. Control Engi. Practi. **8**, 165–173 (2000)

Referees

Stefan Jan Kowalski

Published online: 21 October 2006
© Springer Science+Business Media B.V. 2006

The Guest Editor expresses his cordial thanks to all Referees listed below, who helped review the articles included in this Special Issue of *Transport in Porous Media*, entitled **Drying of Porous Materials**:

Prof. Mieczysław Cieszko
Prof. Yoshinori Itaya
Prof. Mariusz Kaczmarek
Prof. Andrzej Kmieć
Prof. Antoni Kozioł
Prof. Józef Kubik
Prof. Piotr P. Lewicki
Dr. Thomas Metzger
Assoc. Prof. Grzegorz Musielak
Prof. Józef Nastaj
Dr. Wiesław Olek
Prof. Zdzisław Pakowski
Prof. Patrick Perre
Assoc. Prof. Zbigniew Ranachowski
Dr. Jan Stawczyk
Prof. Czesław Strumiłło
Prof. Zbigniew Szwast
Prof. Ryszard Uklejewski
Prof. Jerzy Weres
Prof. Ireneusz Zbiciński

Thank you very much for your time and assistance.

S.J. Kowalski (✉)
Poznan University of Technology, Institute of Technology and Chemical Engineering,
pl. Marii Sklodowskiej Curie 2, 60–965 Poznan,
Poland
e-mail: stefan.j.kowalski@put.poznan.pl